BASIC CONCEPTS AND PASSIVE COMPONENTS

Second Edition

Books by Sy Levine

A LIBRARY ON BASIC ELECTRONICS

Volume One - Basic Concepts and Passive Components
Volume Two - Discrete Semiconductors and Optoelectronics
Volume Three - Integrated Circuits and Computer Concepts

A LIBRARY ON BASIC ELECTRONICS

VOLUME ONE

BASIC CONCEPTS AND PASSIVE COMPONENTS

SY LEVINE

Second Edition

Second Edition

Library of Congress Catalog Card Number: 85-071203
ISBN 0-939-527-006

Portions of this book have previously appeared in the column, TECH TALK, by Sy Levine, published by ELECTRONIC BUYERS NEWS - CMP PUBLICATIONS, INC., and in a series of eight articles that appeared in ELECTRONIC BUSINESS, published by CAHNERS PUBLISHING COMPANY under the title BASIC ELECTRONICS COURSE - FROM OHM'S LAW TO MICROPROCESSORS, by Sy Levine, and are being used with prior agreement.

Printed and bound in the United States of America

Published by:

ELECTRO-HORIZONS PUBLICATIONS
114 Lincoln Road East
Plainview, New York 11803
U.S.A.

ACKNOWLEDGEMENT

The completion of this book would not have been possible without the efforts of two very special people.

To my colleague, Jerry Worthing, whose meaningful suggestions and technical editing were invaluable, I extend my sincerest thanks.

To my wife, Esther, for her support, creativity, commitment, and endless hours of untiring editing, my gratitude and most profound appreciation for all her efforts.

SL

PREFACE

Within the rapidly expanding industrial environment, opportunities are often provided for design and systems engineers to keep in touch with the latest scientific developments. Generally, similar opportunities are not available for engineering-support personnel who need to receive ongoing training and information about new scientific developments. As a result, a communications gap evolves between those with technical knowledge and those who need to learn.

A LIBRARY ON BASIC ELECTRONICS is a three-volume series of books written for the men and women in industry whose need for practical knowledge of electronics has become increasingly timely. The books are dedicated to those who experience the impact of technology and are motivated to explore its intricacies. People in purchasing, sales, marketing, production, quality control, advertising, administration, and management will find this information particularly beneficial.

The books have been specifically designed to provide a comprehensive course of study dealing with the language and content of electronics. The intent of this series is to enable the reader to become familiar with basic concepts and to develop the ability to use the material with a sense of confidence.

BASIC CONCEPTS AND PASSIVE COMPONENTS, Volume One, provides a study of basic electronic concepts, terminology, definitions, and rules. In a step-by-step progression, these principles are developed to establish a firm foundation in the structure of electronic technology. Passive components are examined with regard to how they are made, how they are specified on data sheets, and how they relate to each other in electronic circuitry.

DISCRETE SEMICONDUCTORS AND OPTOELECTRONICS, Volume Two, continues the study of basic electronics by examining the active components of electronic circuits. In a building-block approach, semiconductors and optoelectronic components are examined to provide a more comprehensive picture of the technology and its related state-of-art developments.

INTEGRATED CIRCUITS AND COMPUTER CONCEPTS, Volume Three, examines the technologies of both hybrid and monolithic integrated circuits (the microchip), with regard to how they are manufactured, how they are packaged, their features, and where they are used. The world of digital circuit techniques and basic computer concepts is explored through the study of binary arithmetic (the language of the computer), logic circuits, memories, and the other elements of the modern computer.

Each book in the series is enhanced with illustrations, glossaries, reinforcement exercises, and a comprehensive index. In addition, future trends in all phases of the technology are included.

It is the author's hope that A LIBRARY ON BASIC ELECTRONICS will provide the basic information and insight into electronics technology and offer the reader an opportunity for greater productivity and gratification in his or her work.

TABLE OF CONTENTS

A BRIEF HISTORY OF ELECTRONICS

"What is past is prologue"
The Tempest - William Shakespeare

To the ancients, the world was full of magic and mystery. The sunrise and sunset, the change of seasons, an eclipse of the moon, and a bolt of lightning were among the many wonders of the physical universe which were regarded with fear and reverence. Since no organized system existed to study natural phenomena, these awe-inspiring occurrences were perceived as manifestations of supernatural power and not as forces governed by natural laws. The marvels of the universe were yet to be examined and classified to establish a base of scientific data.

Eons passed before the laws of physics and chemistry were developed and natural phenomena were scientifically explored. Although limitations were maintained by the inadequacy of available instrumentation and materials, new sciences continued to evolve while more and more information was being accumulated. The scope of research continued to expand despite the resistance in the scientific community to new ideas and to changes in the established order. In many cases, acceptance of innovations came years after their introduction - sometimes centuries later.

Through the persistence of inventive minds, new ideas were probed and new sciences were established. During the 18th, 19th, and 20th centuries, dramatic developments in the rapidly evolving sciences occurred on a wide international scale. This continuing process eventually led to the study of the principles of electricity, and later, by combining this ground-breaking data with other emerging sciences, the technology of *electronics* was created.

Today, many of the people who created and developed these sciences are honored for their outstanding achievements, and their names have become an integral part of the current scientific vocabulary. Although there have been countless numbers of scientists who have made noteworthy contributions, the following brief history highlights a few.

Formal investigation of electrical phenomena is considered to have begun in 1752, when **Benjamin Franklin** flew a kite during a thunderstorm to investigate the properties of lightning. The results of this famous experiment helped answer some questions about lightning and its effects which had intrigued the human mind for centuries. With his pioneering research efforts, Franklin established many basic principles and theories dealing with positive and negative electrical charges.

James Watt, a Scottish engineer, invented a practical *steam engine* in 1769. It became the first machine using steam as a readily available source of energy to provide efficient mechanical rotation. Sixty-two years later, **Michael Faraday**, an English physicist, chemist, and mathematician, expanded on the concept of the steam engine by developing the first *electrical generator*. This invention has had a profound effect on all future scientific work in the field of electricity.

In 1778, **Alessandro Volta**, an Italian physicist, created a device that was used to store electrical charges and, eventually, it became the basis for the modern *capacitor*. In 1800, Volta reached a major goal of his life's efforts with the invention of the *electric battery*.

In 1820, **Hans Christian Oersted**, a Danish physicist, founded the study of *electromagnetism* by demonstrating that a significant relationship existed between electricity and magnetism. This discovery was a turning point in magnetic and electrical theory and initiated new investigations into the field of electromagnetism by scientists in other countries.

André Ampère, a French physicist, studied *electrical currents* in motion and established some of the basic laws describing these phenomena. In 1823, he identified the existence of static electricity and introduced the term *electrostatics* to the language. His theories expanded the existing knowledge of magnetic energy and were used to further the work on electrical charges which Franklin had begun.

Georg Simon Ohm, a German physicist, identified a fundamental electrical relationship in 1827 that was developed through empirical laboratory research. He discovered that a connection existed between electrical *pressure (voltage)*, electrical *resistance*, and electrical *current*. Ohm defined this relationship as a straightforward mathematical equation that came to be known as *Ohm's Law*. Although this law is quite simple, it has become an extremely profound concept since it is the basis of all electrical and electronic circuit design.

Nearly fifty years earlier, **Henry Cavendish**, an English physicist, had recognized the same relationship between the electrical parameters that would later be investigated by Ohm. Cavendish, however, had failed to publish his findings and, as a result, was not officially acknowledged for his work. It was Ohm who was given credit for developing this basic law of electricity several years after the publication of his own independent research.

Through his work on magnetic forces in a wire, **Michael Faraday** developed the first theories on *magnetic induction* leading to the invention of the *transformer* in 1822. Similar independent work was being done on the *properties of magnetism* at about the same time by **Joseph Henry**, an American physicist. Although neither Faraday nor Henry initially realized the full implication of their work, their research led to significant developments in the field of electricity. Faraday also recognized the relationship between chemistry and electricity and, in 1832, established the rules describing the flow of *electrical current in chemicals*, known as *Faraday's Laws.*

Alexander Graham Bell, a Scottish engineer who emigrated to the United States, invented the *telephone* in 1876. This invention was based on the theories of the American physicist, Joseph Henry and a German physiologist and physicist, **Hermann von Helmholtz,** who studied anatomy with emphasis on the structure of the ear.

In 1877, **Thomas Alva Edison**, America's most prolific inventor, improved the design of Bell's telephone and turned it into a commercially practical instrument. In 1876, Edison set up the first industrial research laboratory in Menlo Park, New Jersey and initiated the concept of the research team.

In 1879, Edison produced a commercially practical *incandescent electric lamp* that is considered to be his greatest contribution. Edison played a key role in establishing the concept and design of the first direct current (DC) power station in the United States. It was built and installed in New York City in 1881 and made the electric incandescent lamp available for general use.

Among his other notable inventions were the *carbon microphone*, the *phonograph*, the *movie projector*, and a combination of these two innovative machines which he called the *kinetophone*. Many years later, the kinetophone became the basis of a new cinematic technology called the "talkies" and a dramatic new direction in the nature of film production was initiated. He patented over 1300 inventions during his lifetime, a record that has never been equaled. Although Edison's work was more practical than theoretical, in 1883, while attempting to improve the electric lamp, he made an important scientific observation. He found that electrons flowing from a heated element were able to move through a vacuum onto another metallic surface in a circuit. This process was later called the *Edison effect* and was patented in 1884. Since he believed that there was no immediate application for this phenomenon, Edison merely described it in the technical literature but did no further work on it.

In 1903, a more theoretical approach by **Sir John Fleming**, an English physicist, applied the Edison effect to create a unique electrical device called an *electric valve* . Because of its unidirectional feature, current was forced to flow in only one direction. In the United States, it was referred to as a *vacuum tube diode* and was applied in circuitry to change alternating current (AC) to direct current (DC).

Nikola Tesla, a young scientist from Croatia, began his career as an inventor in Austria and emigrated to the United States in 1884. Despite the fact that his work was generally unacknowledged by his contemporaries, his contributions were outstanding. He developed the technique for generating alternating current and, through his efforts, the first AC power station was built at Niagara Falls, New York. Tesla investigated the science of *radar* long before it was fully developed by the British during World War II. He also explored the science of *automation*, the forerunner of *robotics*.

Tesla's work on *wireless communication* (radio) brought him into conflict with **Guglielmo Marconi**, an Italian engineer, who obtained the first patent for wireless communication in 1896. In 1943, the U.S. Supreme Court acknowledged Tesla's prior work on this technology and ruled that the patent was to be taken from Marconi and awarded to Tesla. Nicola Tesla, however, had died in poverty six months previously and, unfortunately, was only acknowledged posthumously for his pioneering work.

Charles Steinmetz, a German engineer, was universally recognized as one of the world's foremost geniuses in the field of electrical research. He fled to the United States in 1889 because of religious and political persecution and, four years later, became the technical genius behind the rapid growth of the General Electric Company. His work emphasized all aspects of *alternating current* (AC), and he patented over two hundred inventions in this field. Though the initial work was done by Tesla, AC power became the dominant source of electrical energy throughout the world because of the continuing efforts of Steinmetz.

In 1906, **Lee De Forest**, an American inventor, added another element to the vacuum tube diode to provide the capability of electrical amplification. Although he initially called the device an *audion*, it became known as a *vacuum tube triode*. Before the advent of the triode, amplification of sound by electrical means had not been possible. De Forest's work opened the way to the development of radio, television, and eventually, the computer. This achievement was the giant step that vaulted electrical circuitry into the next technological generation - *electronics*.

In 1918, **Edwin Armstrong**, an American electrical engineer, devised an innovative tuning approach for use in radio receivers. It was called a *superheterodyne system* and provided an easy, quick, and precise way of tuning to a specific radio station. This technique changed the nature of radio communications and eventually contributed to the wide acceptance of commercial radio. In 1939, Armstrong developed the concept of *frequency modulation* (FM), a radio transmission technique that virtually eliminated interference in the transmission and reception of audio information.

Electronic engineers were now creating new circuit designs and restructuring existing systems with the use of a variety of different types of vacuum tubes and a wide variety of other electronic devices, called *passive components*. Among these were: *fixed and variable resistors, capacitors, mechanical switches, relays, crystals, inductors, and transformers*. Improvements were being made in the electrical characteristics and structure of many of these components. Greater capability and increased reliability offered improved system performance and encouraged newer applications.

In the 1940s and early 1950s, the efforts and creativity of countless design engineers led to the development and production of such electronic innovations as radar, commercial television, sophisticated instrumentation systems, and control systems for the industrial and medical fields.

A major advancement in the manufacturing technique of electronic equipment was initiated in the early 1940s with the introduction of the *printed circuit* (PC) *board*, a component assembly and packaging process that consisted of soldering and interconnecting individual components on a thin, flat, pre-wired insulating board. Production time and costs were appreciably reduced and testing, maintenance, and replacement procedures were simplified and made more flexible by this new *modular* approach.

During this period, the sciences of chemistry, metallurgy, and photolithography were combined in an intensive research program to produce *solid-state devices* called *semiconductors*. In 1948, three American physicists at Bell Labs in New Jersey, **John Bardeen, Walter Brattain**, and **William Shockley**, collaborated to invent a revolutionary new semiconductor called a *transistor*. It was an extremely small, solid-state version of the vacuum tube triode and was used to control and amplify current in a circuit using very little power. In 1956, the three scientists were honored for their work and were awarded the Nobel Prize in Physics.

In 1954, the semiconductor industry was born. Transistors, semiconductor diodes, and other solid-state devices were made available for military, space, commercial, industrial, and consumer applications. At the end of the 1950s, and into the 1960s, semiconductors were replacing vacuum tubes by retrofitting existing circuits and were also being used to create many new circuit applications.

Semiconductor technology led to remarkable new approaches in design, packaging techniques, and most important of all, improved system reliability. In addition, the need for electrical power in electronic systems was greatly reduced and the miniaturization of circuitry and equipment was made possible.

In 1958, **Jack Kilby**, an American physicist at Texas Instruments in Dallas, Texas, moved the concept of the PC board a long leap forward by devising a means of constructing a complete circuit on a single, thin germanium wafer. Advances in metallurgy, chemistry, and photolithography provided the capability of creating and interconnecting all the required circuit components on that germanium wafer. This structure was called an *integrated circuit* (IC), however, it was not a commercial success because of inherenty problems in mass production of the device. Despite this setback, Kilby's idea introduced the revolutionary concept of integrated circuit technology.

By the late 1950s, semiconductors had gone through some major processing changes. Although germanium had some use in isolated applications, silicon had replaced it as the preferred semiconductor material because of its many electrical and thermal advantages.

In 1959, **Jean Hoerni**, a Swiss physicist, and **Robert Noyce**, an American physicist, at Fairchild Semiconductor in Mountain View, California, used silicon as the base material to develop the *planar process*, a stable technique for the production of semiconductors. This innovative process led to the creation of the first commercially successful integrated circuit that contained all component functions within a single silicon chip. This new chip was created by a production method called *monolithic IC technology.*

In 1961, the first ICs were mass produced and were followed by the industry's overwhelming acceptance of more complex integrated circuits. From 1962 to the present, further improvements in production processes expanded this science to provide highly concentrated circuits and systems on an extremely small silicon chip.

As of this writing, integrated circuits are being manufactured with more than 64,000,000 component functions on a ¼ inch by ¼ inch chip of silicon. Component functions are metallurgically interconnected on a chip to form a circuit, many circuits, a system, or many systems that can be manufactured at relatively low cost. This processing technique is referred to as *Ultra Large Scale Integration* (ULSI)) and has made available many applications that previously had been impossible to design.

The age of *microcomputers* began in 1971 when the Intel Corporation produced the first *microprocessor* chip. The introduction and use of microprocessor/microcomputer technology has had a major and dramatic impact on world-wide science and industry, bringing with it the development and introduction of the *personal computer* (PC). Prior to this development, use of computers were limited to large company labs and research centers because of their extremely large size and high cost. Now small enough to be held on a lap, or in the palm of one's hand, the PC has brought this spectacular technology into private homes and small offices at relatively low cost.

The advancement of electronics technology is based on a continuity of creative thinking coupled with intensive research. From early investigation into the effects of lightning, to the study of electromagnetism and associated sciences, to the development of new circuit concepts and the availability of components and materials to implement these concepts, electronic technology has progressed to a point of sophistication that has produced astounding results.

The lack of scientific knowledge during ancient times has evolved into the modern world of electronics, illuminating new landscapes and encouraging adventures into the unknown. The extraordinary monolithic integrated circuit, the modern computer, and the related technologies herald an exciting future of infinite possibilities.

PART ONE
BASIC CONCEPTS

1 CHAPTER ONE

PRODUCING ELECTRICAL ENERGY

CIRCUITS - ELECTRICAL AND ELECTRONIC

ENERGY CONVERSION TECHNIQUES

- MECHANICAL ENERGY TO ELECTRICAL ENERGY
- CHEMICAL ENERGY TO ELECTRICAL ENERGY
- SOLAR ENERGY TO ELECTRICAL ENERGY

PRODUCING ELECTRICAL ENERGY

CIRCUITS - ELECTRICAL AND ELECTRONIC

A **circuit** is an unbroken electrical path containing a single component, or a combination of several components, connected to a source of electrical energy. Electrical energy, or *electricity,* is the force that provides the power to make circuitry function.

The terms *electrical* and *electronic* refer to two types of circuits designed to function in different ways. Despite their differences, both types of circuits are based on the same laws of physics.

All appliances and equipment requiring electrical energy to operate have the following characteristics in common:

- They are enclosed in a package containing all parts of a system, including mechanical, electrical, and/or electronic devices.
- They include a means of accessing an external or internal source of electrical energy to provide the power for operation.
- They are operated by that part of the assembly called the circuit .

Electrical circuitry is structured to operate simple control systems where *electrical amplification* (signal enlargement) is not required. Reflecting the earlier of the two technologies, electrical circuits are typically used in simple lighting systems, refrigerators, home appliances, air-conditioners, hair dryers, toasters, coffee machines, and similar circuitry that could be part of more complex systems that also contain some electronic circuits.

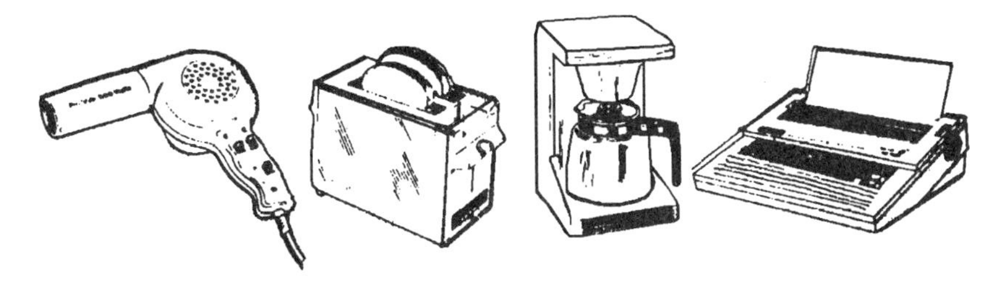

Equipment with Electrical Circuitry
Figure 1.1

Electronic circuitry, a more advanced technology, is used to provide intricate switching functions and amplification of electrical signals. Although both electrical and electronic circuits may use complex components and can be interconnected in a complicated manner, only electronic circuits are capable of electrical amplification.

> **Electrical amplification** is the process that acts on an electrical signal at the input of an electrical amplifier to produce an output that is greater than the input signal.

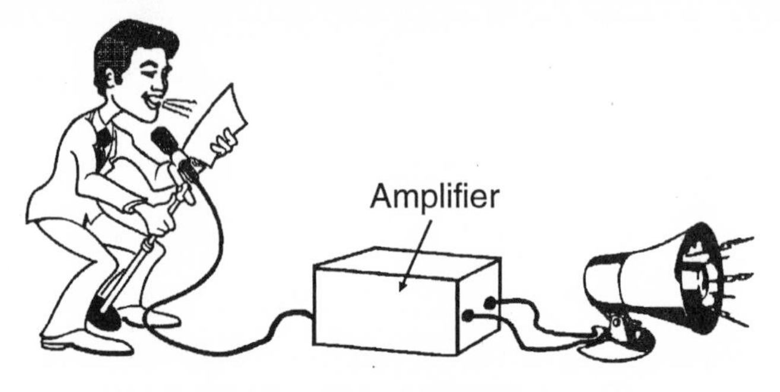

Electrical Amplification
Figure 1.2

Electronic circuits are used in radio, television, audio systems, communications systems, sophisticated medical diagnostic equipment, calculators, and computers. In recent years, traditional electrical systems have been made more functional and reliable with the addition of electronic circuits.

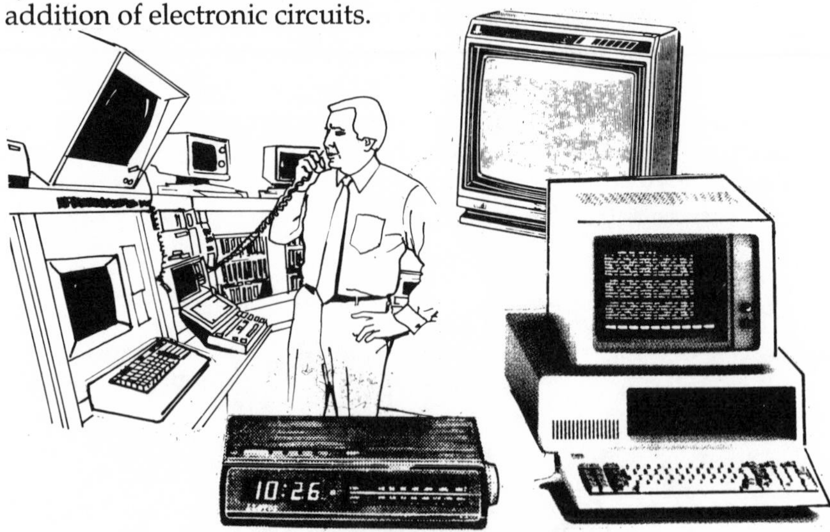

Equipment with Electronic Circuitry
Figure 1.3

ENERGY CONVERSION TECHNIQUES

Electrical energy is not naturally available as usable power and must be created from *mechanical, chemical*, and *solar* energy sources that exist freely in nature. When electrical appliances, machines, and equipment are working properly the source of their power is taken for granted. When power is lost for any length of time, there is a sudden awareness that electrical energy has great significance. Restoring the power quickly becomes urgent while many questions arise regarding reasons for the power loss.

Understanding the origins of electrical energy and how it is generated will explain how existing natural forces are harnessed to provide a source of electrical power. To change the different forms of natural energy into usable electricity, several techniques of conversion must be used.

MECHANICAL ENERGY TO ELECTRICAL ENERGY

Early scientific researchers discovered that an invisible field, called *magnetism*, exists across the poles of a magnet. When a wire is moved through this field, electrical energy (called the *electromotive force*, or EMF) is generated at the ends (terminals) of the wire. Electrical energy is produced by using an *electrical generator.*

An electrical generator consists of one or more electrically connected coils called an *armature*, a magnet mounted around the armature, and a means of rotating the armature within the magnetic field. Mechanical energy provides the force to rotate the armature within the magnetic field to produce an electromotive force at the terminals of the generator. (See Figure 1.4). The EMF at the terminals can be increased by:

- Increasing the number of turns of wire to create a larger coil
- Winding the coil around a core of iron (or any ferrous material)
- Using a stronger magnet
- Combining any of the above

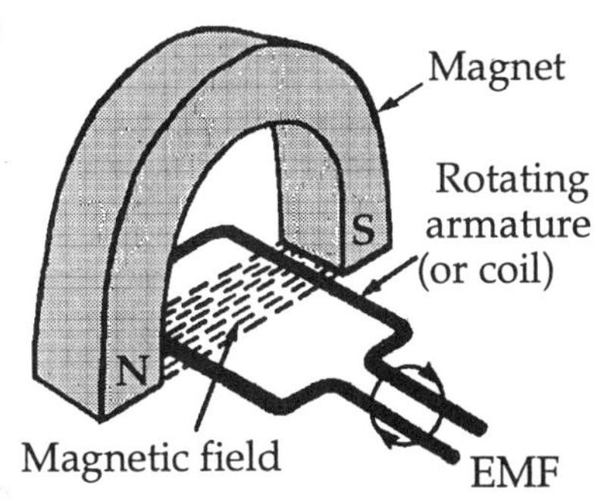

Electrical Generator Concept
Figure 1.4

Three forms of mechanical energy can be used to provide the required rotation of an armature.

Steam power - When water boils, steam is produced. As water continues to boil, steam expands to provide mechanical energy that rotates an armature in an electrical generator. (See Figure 1.5)

The energy needed to boil water to produce steam is provided by:

- Burning fossil fuels such as: coal, oil, wood, or natural gas
- Collecting the heat of the sun's rays
- Splitting of nuclear material

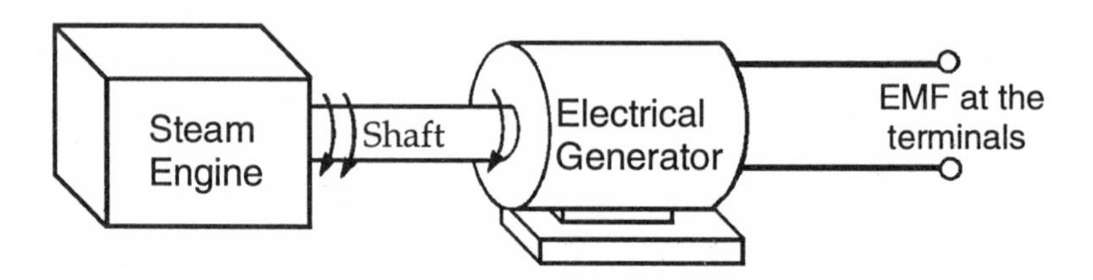

Mechanical Conversion by Steam Power
Figure 1.5

Water power - The movement of water flowing over turbine blades causes the rotation of an electrical generator. (See Figure 1.6)

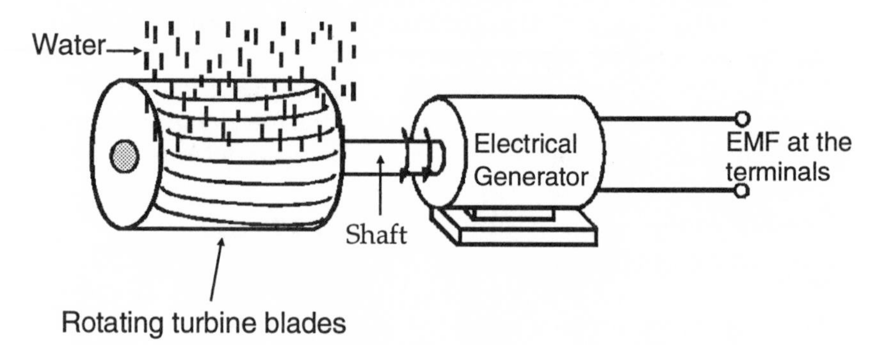

Mechanical Conversion by Water Power
Figure 1.6

A turbine consists of a series of blades coupled to the shaft of a generator. Placing the blades in the path of flowing water will cause the blades to rotate. The armature in the generator's magnetic field will rotate, generating an electromotive force at its terminals.

Wind power - The movement of air across the rotatable blades of a carefully balanced windmill will cause rotation of an armature in the magnetic field of an electrical generator, creating an EMF at the terminals of the generator.

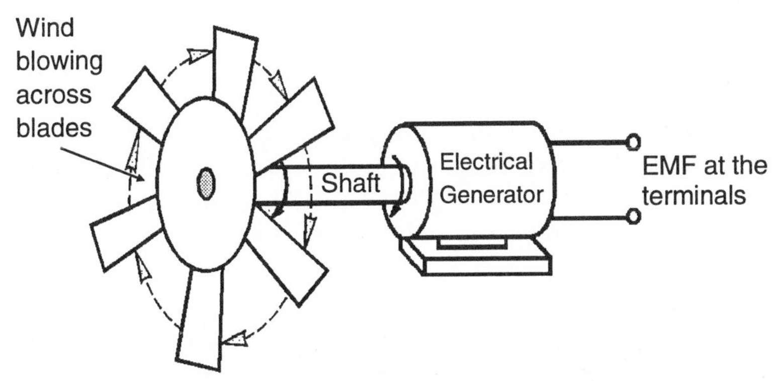

Mechanical Conversion by Wind Power
Figure 1.7

CHEMICAL ENERGY TO ELECTRICAL ENERGY

Electrical energy (EMF) can be produced through chemical means, the method by which battery power is created. Two metallic strips, (*electrodes*), are inserted in a solution of specific chemicals (*electrolyte*). The chemical action between the electrolyte and the electrodes in a container produces an electromotive force at the external terminals connected to the electrodes.

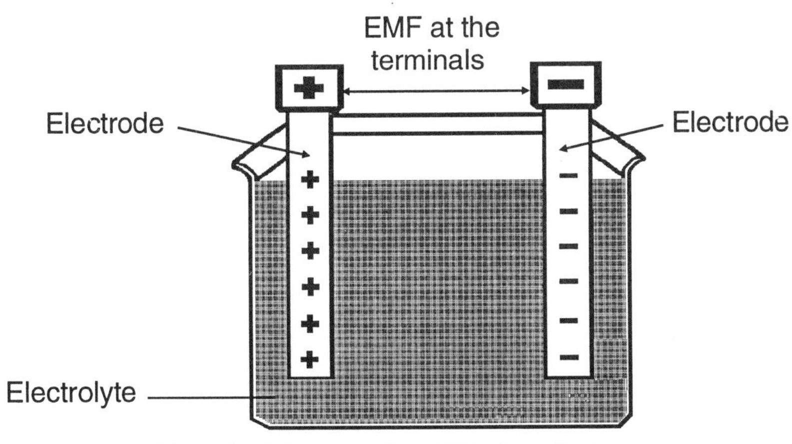

Chemical Conversion Within a Battery
Figure 1.8

SOLAR ENERGY TO ELECTRICAL ENERGY

The sun is the source of limitless solar energy in the form of light and heat. As a source of heat, solar energy can be used to change water into steam to convert mechanical energy into electricity. Another approach, however, is to convert the light of the sun directly into electrical energy without any intermediate steps.

When the rays of the sun strike certain light sensitive materials, they will exhibit a phenomenon called the *photovoltaic effect* - the conversion of light directly into electricity. The photovoltaic effect can be achieved with a variety of materials, including silicon, selenium, cadmium sulfide, germanium, gallium arsenide, and amorphous glass.

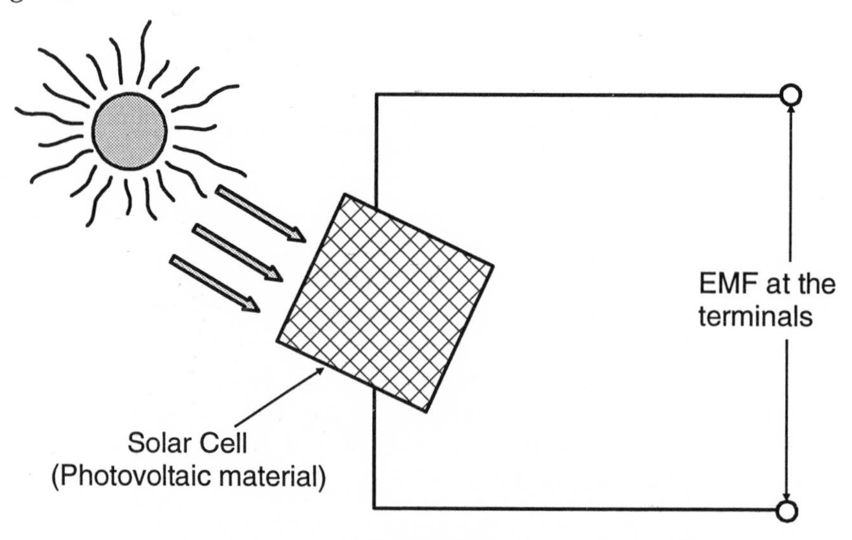

Solar Conversion With Photovoltaic Material
Figure 1.9

The criteria for choosing any particular technique to convert any form of energy to electricity depend on conversion efficiency, cost, availability of materials, environmental conditions, the time needed for installation, as well as legal and political restraints.

Once electricity is generated by any of the above techniques, it becomes available as the power needed to operate electrical and electronic circuits and systems. Without it, the equipment in homes, offices, plants, automobiles, in fact, anything that requires electrical energy for its operation would stop functioning.

2 CHAPTER TWO

WATER ANALOGY

COMPARISON OF A WATER SYSTEM WITH AN ELECTRICAL CIRCUIT

- WATER FLOW FROM A WATER TOWER
- WATER FLOW THROUGH A FISH TANK
- ELECTRICAL CURRENT FLOW

WATER ANALOGY

COMPARISON OF A WATER SYSTEM WITH AN ELECTRICAL CIRCUIT

In its basic form, a circuit consists of a single electrical component connected to a source of electrical energy. When an electrical circuit is in the form of a continous loop, it is called a *closed circuit* and current will flow through it. An electrical circuit is similar to a closed water system because the flow of electrical current through a circuit is comparable to the flow of water through a hose. The water analogy is a classical method of illustrating the operation of an electrical circuit. (See Figure 2.1)

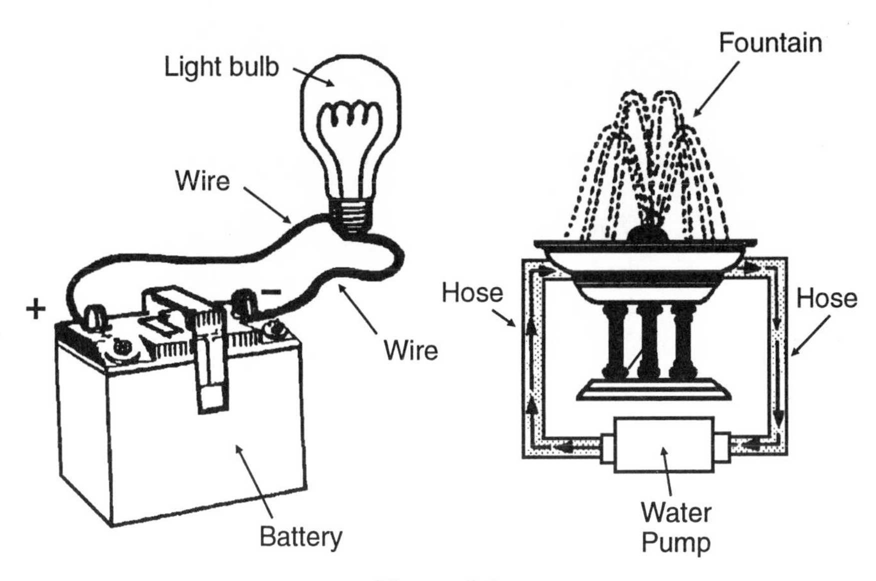

Figure 2.1

WATER FLOW FROM A WATER TOWER

In Figure 2.2, a water tower is standing on stilts at a height of 50 feet. The tower can be filled by rainfall or by having water pumped in from any external source. The volume of water in the tower has an inherent downward force, or push, produced by the force of gravity. This push, called *water pressure*, is measured in pounds per square inch (psi). To move the water from the filled tower, a valve is connected at the base of the tower. The water pressure in the tower remains constant if the height of the tower remains the same.

With the valve fully opened, there is very little *resistance*, (opposition) to the water pressure and a large number of gallons per minute will flow from the tower through the valve.

The rate of flow is called *water current* and is measured by the number of gallons flowing from the valve in one minute (gallons per minute or gpm).

With a change in the height of the tower, and the valve at its fully-open setting, the water pressure will change, producing a corresponding change in the rate of flow of the water. If the tower is raised, water pressure will increase and water current will also increase. (See Figure 2.3) If the tower is lowered, the water pressure will decrease and water current will also decrease.

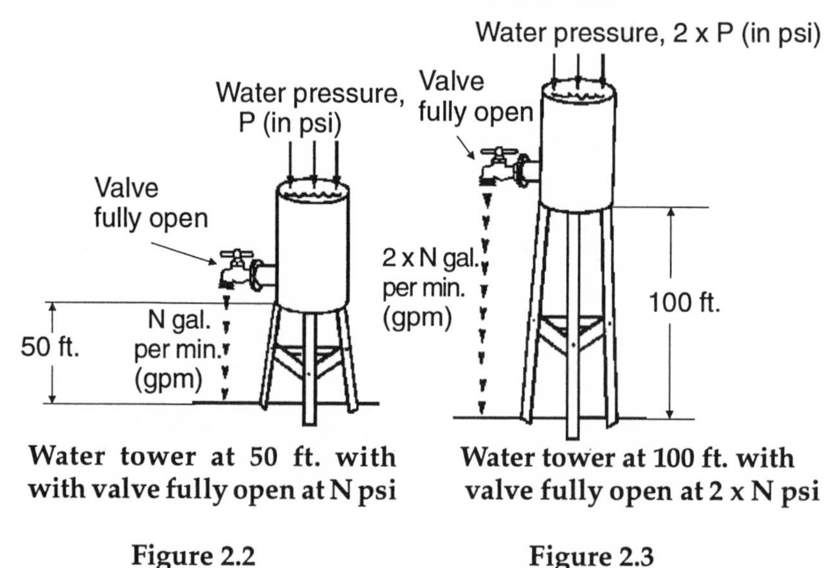

Water tower at 50 ft. with with valve fully open at N psi

Figure 2.2

Water tower at 100 ft. with valve fully open at 2 x N psi

Figure 2.3

The change in water current is *directly related* to the change in water pressure. Resistance to water pressure depends on the setting of the valve and is constant when the valve setting is fixed.

The water pressure remains at a constant value when the tower is at a constant height. When the valve is partially closed, there is an increase in resistance to the water pressure, resulting in a decrease in water current.

A further closing of the valve results in a further increase in its resistance to the water pressure, resulting in a further decrease in water current. (See Figures 2.4 and 2.5)

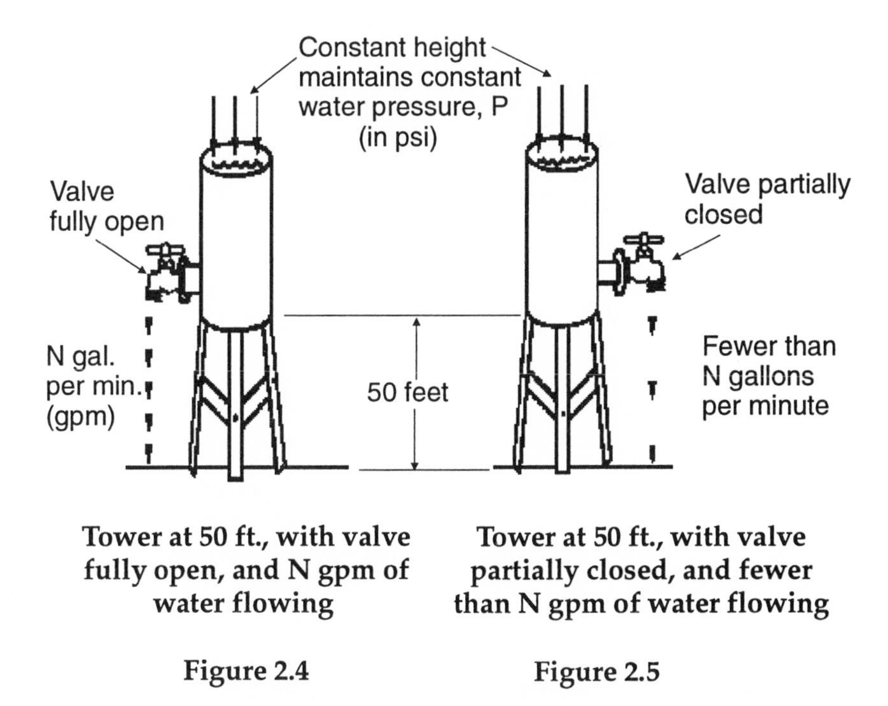

Tower at 50 ft., with valve fully open, and N gpm of water flowing

Figure 2.4

Tower at 50 ft., with valve partially closed, and fewer than N gpm of water flowing

Figure 2.5

With water pressure held constant, changes in water current are *inversely related* to changes in resistance to the water pressure. The resistance of the valve depends on its setting.

- Increased valve resistance will cause a decrease in water current.
- Decreased valve resistance will cause an increase in water current.

WATER FLOW THROUGH A FISH TANK

The example of a variable pressure pump connected to a fish tank is another illustration of the water analogy. (See Figure 2.6) When the pump is turned on, water is pushed from the exit side of the pump through a hose (1) into the fish tank. Water leaves the fish tank through a second hose (2) and returns to the other side of the pump.

As long as the pump is kept at a constant pressure and there is no change in the diameter or condition of the hoses, the water will keep recirculating at a constant rate of flow.

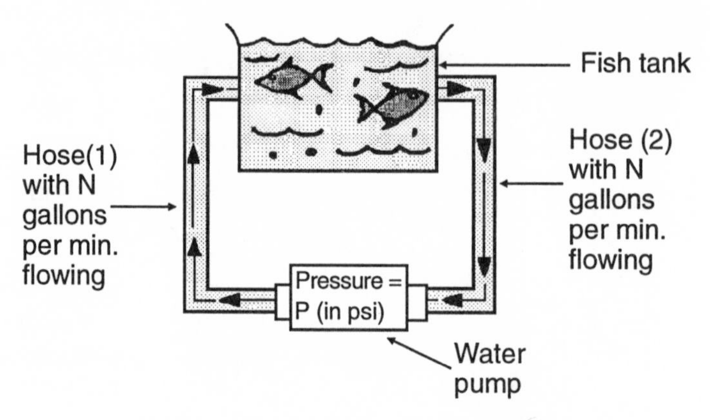

Fish tank with pump pushing water through hose (1) and returning water to other side of pump through hose (2)

Figure 2.6

If the water pressure of the pump is increased without any change in the hoses, water current will also increase. (See Figure 2.7) If the water pressure is decreased, water current will also decrease.

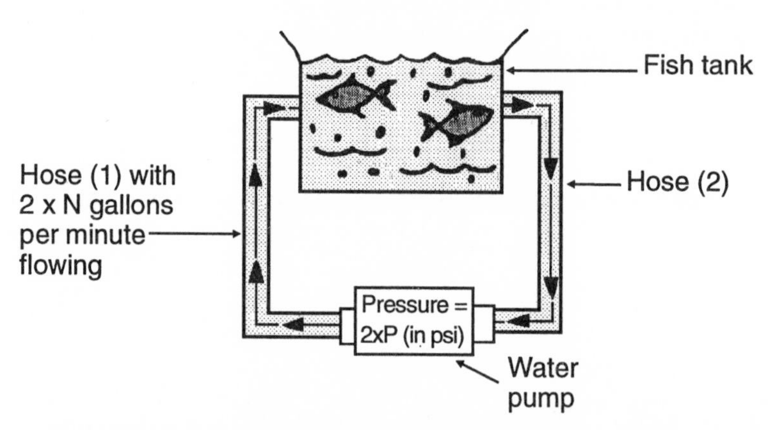

Fish tank with pump operating at twice the previous water pressure resulting in twice the flow of water current

Figure 2.7

A change in the flow of water current is *directly related* to a change in water pressure. Any dirt or debris inside a hose will reduce the inside diameter of the hose, increasing its resistance to the water pressure. With water pressure held constant, the result will be a decrease in the flow of water current. If a hose becomes completely blocked, the flow of water totally stops. Since the resistance of the hose is essentially infinite (although water pressure is still present), no water is flowing in the system.

With water pressure remaining constant, a change in water current is *inversely related* to a change in the resistance of the system. (See Figures 2.8 and 2.9)

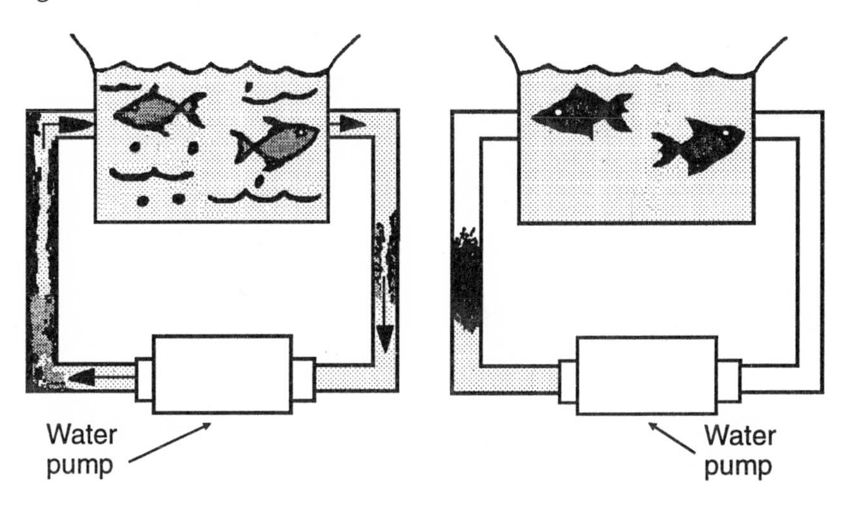

Fish tank with pump operating and some debris in hoses, decreasing water current in the system

Figure 2.8

Fish tank with pump operating but debris in hose totally stopping water current in the system

Figure 2.9

A water system operates as a closed system, or a *closed loop*, which is essential to keep the system operating properly. In contrast, a system that is not closed indicates that there is a gap between parts of the system and the system is no longer functioning. This condition is referred to as an open or non-operating system.

In a similar manner, an electrical or electronic circuit operates as a closed loop, or closed circuit, which is essential to keep the circuit operating properly. If there are gaps in a circuit, connections between parts of the circuit have been broken, or disconnected, and the circuit is no longer operating. This condition is called an open or non-operating circuit.

ELECTRICAL CURRENT FLOW

In an electrical or electronic circuit, the relationship between electrical pressure, electrical resistance, and the resulting flow of electrical current is analogous to the relationship between water pressure, resistance of a hose to the water pressure, and the resulting water current. In an electrical or electronic circuit, resistance is an inherent characteristic of the components that make up the circuit.

With resistance in a circuit remaining constant, changes in current will vary *directly* with changes in pressure.

With RESISTANCE remaining constant:

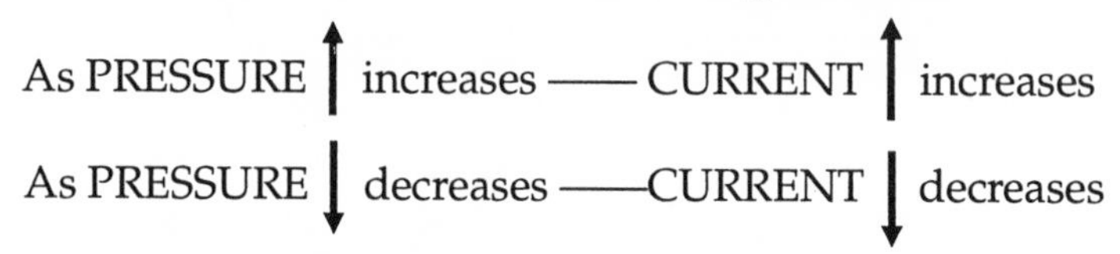

As PRESSURE increases —— CURRENT increases
As PRESSURE decreases ——CURRENT decreases

With pressure remaining constant, changes in current vary *inversely* with changes in resistance.

With PRESSURE remaining constant:

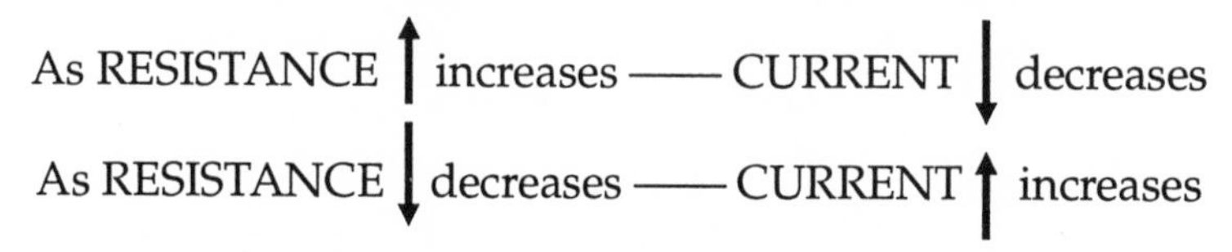

As RESISTANCE increases —— CURRENT decreases
As RESISTANCE decreases —— CURRENT increases

In a manner that is analogous to the fish tank water system, an electrical or electronic circuit will stop operating if the resistance in the circuit is increased to infinity, creating an open circuit. The current will stop flowing even though the pressure is still present. This condition of infinite resistance in a circuit can be caused by a break or burn-out in a connecting part of the circuit, a defective component, or by the circuit being deliberately switched OFF.

In both electrical and electronic circuits, the specific relationship between the **electrical pressure**, the **electrical resistance**, and the resultant flow of **electrical current** is known as **Ohm's Law**. This fundamental relationship is the basis for all electrical and electronic circuit design. A definitive discussion of Ohm's Law is covered in Chapter Four - Ohm's Law.

3 CHAPTER THREE

ELEMENTS OF A CIRCUIT

VOLTAGE

RESISTANCE

CURRENT

CONDUCTORS

METERS

GRAPHIC SYMBOLS

SUBSCRIPT AND SUPERSCRIPT NOTATIONS

Chapter Three

ELEMENTS OF A CIRCUIT

VOLTAGE

Voltage is the electromotive force (**EMF**) that pushes current through an unbroken electrical path, called a *closed circuit*. The **volt** is the unit of measurement for voltage.

- Electromotive force is analogous to the water pressure used in the water analogy.
- When used as the source of electrical pressure, voltage is the electrical force produced by a battery, laboratory voltage supply, generator, or solar cell. It can be referred to as the *supply voltage, voltage supply, source voltage,* or *voltage source* and is designated by the letter **E**.
- The voltage created across the terminals of a component, or across the two points in a circuit to which a component is connected is designated by the letter **V**.
- The voltage that exists at the terminals of a voltage source, **E**, or at the terminals of a component, **V**, is referred to as the voltage *across* or *between* these terminals.

DC VOLTAGE SOURCE

A **DC** (direct current) voltage has a constant *polarity* with one of its terminals is marked **positive**, or plus (+), and the other, **negative**, or minus (-). It is available as either a *fixed* or *variable* DC voltage supply.

The amount of voltage that exists across the terminals of a DC voltage source is called its *magnitude.*

The magnitude of a fixed DC voltage source (as in a battery) is not adjustable and will remain at its specified value as long as the voltage source is not impaired or changed. The magnitude of a variable DC voltage source (as in laboratory test equipment) can be adjusted from zero voltage to a specified maximum value of DC voltage.

A fixed DC voltage source is shown schematically in Figure 3.1 and drawn as a series of long and short bars, with the long bar marked as the positive terminal. In Figure 3.2, an arrow drawn diagonally through the DC voltage symbol indicates a variable supply voltage.

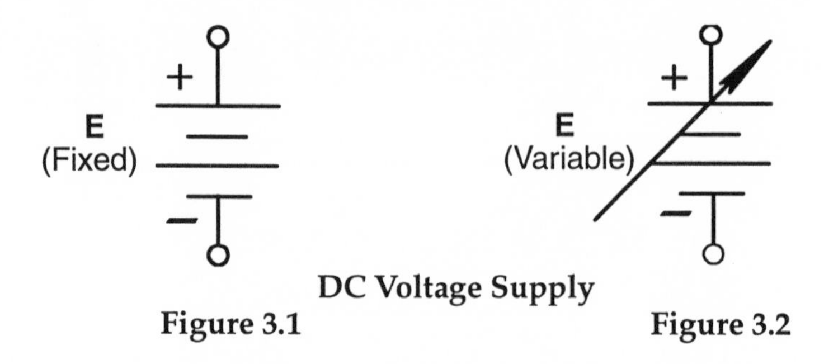

DC Voltage Supply

Figure 3.1 Figure 3.2

AC VOLTAGE SOURCE

An **AC** (alternating current) voltage source, unlike DC, has no polarity designation since its voltage polarity reverses during one half of its cycle. A positive voltage exists during half its cycle and a negative voltage exists during the remaining half cycle.

- The *frequency* inherent in an AC waveshape defines the number of cycles per second.

- The *amplitude* of an AC voltage source is the amount of voltage across the voltage source and can be either fixed or variable.

With the fixed AC supply voltage of Figure 3.3, the amplitude is at a fixed value. The symbol of the variable AC voltage supply is shown in Figure 3.4.

Note that there are no polarity designations at the terminals of an AC supply voltage, since the polarity of an AC voltage is not constant but is continuously alternating.

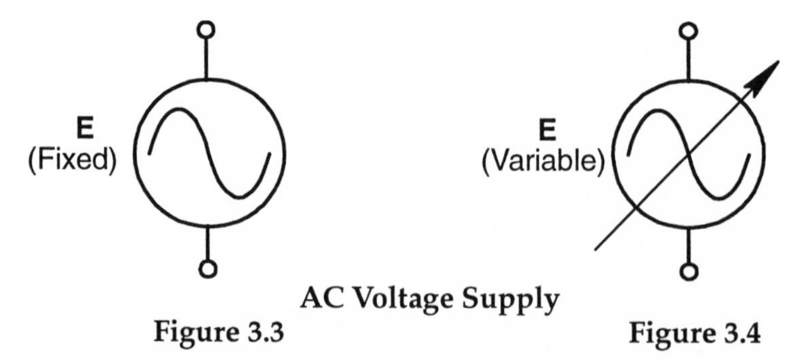

AC Voltage Supply

Figure 3.3 Figure 3.4

A detailed explanation of DC and AC voltages, polarity, sine waves, frequency, magnitude, and amplitude are discussed in Chapter Seven - Alternating and Direct Current.

RESISTANCE

Resistance is the electrical opposition to voltage and is designated with the letter **R**. In conjunction with voltage, resistance determines the amount of current flow in a circuit. The unit of measurement for resistance is the **ohm**. In a circuit diagram,the last letter of the Greek alphabet, **omega** (Ω), is the symbol for ohms .

Electrical resistance is analogous to the physical resistance of the partially closed valve or obstructed hose used in the water analogy. Just as water pressure is used with an adjustable valve to produce a functioning water system, similarly, in an electrical or electronic circuit, electrical pressure (voltage), is used with electrical resistance to produce a functioning circuit. Resistance is inherent in every material with its ohmic value depending on its molecular structure, size, and temperature.

Commercially available fixed and variable resistance components are called **resistors** and are discussed in Chapter Nine - Resistors.

ELECTRICAL LOAD

An *electrical load* can be compared to a physical load. A mechanical force or pressure can be applied to a physical object (load) to move it from one place to another. This physical load is measured in units of weight (ounces, pounds, grams, etc.).

To have an electrical or electronic component do work, an electrical force or pressure (voltage) must be applied to it. The component being energized is an electrical object (load), and its resistance is measured in ohms (Ω). A resistor is an example of a load, as is a light bulb, an audio amplifier, radio, TV set, or a computer.

The resistance of a load can be:

- Measured with an ohmmeter.
- Either printed or color-coded to identify its value and tolerance
- Specified in a manual furnished by the equipment manufacturer
- Calculated by using the appropriate Ohm's Law equation

The graphic symbol for fixed resistance is shown in Figure 3.5. Variable resistance is illustrated with a diagonal arrow drawn through the fixed resistance symbol. (See Figure 3.6) An optional *block symbol* for fixed resistance is shown in Figure 3.7, and is often used to symbolize a fixed load.

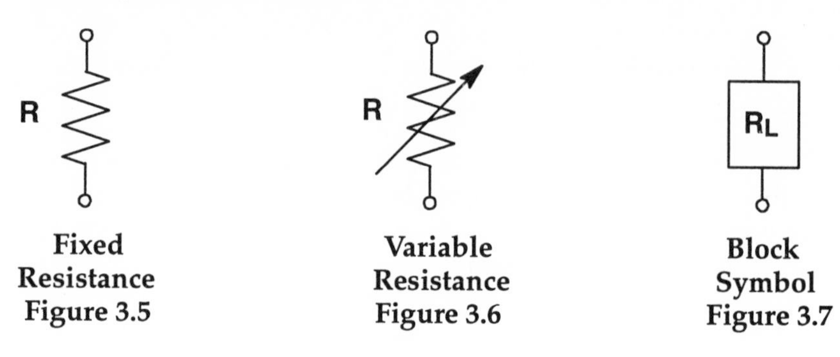

Fixed Resistance Figure 3.5

Variable Resistance Figure 3.6

Block Symbol Figure 3.7

CURRENT

Current refers to the rate of flow of electrons in a circuit and is designated by the letter **I**. The unit of current measurement is the **ampere** (A) or **amp.**

The word **electron** is derived from "elektron", the Greek word for amber. Since ancient times, it has been known that when amber, a resinous material, is rubbed, it exhibits its inherent ability to attract objects that are light in weight.

This unique characteristic of amber and other materials, such as glass, hard rubber, and a variety of gems, was first pointed out in the early 1600s by **William Gilbert**, an English physicist. He categorized all such materials under the name of *electrics*, the term he derived from the word elektron.

Electrical current is indicated by the measurement of a large number of electrons (the smallest unit of electrical charge) passing any point in a circuit in one second. This quantity, 6,250,000,000,000,000,000 (or 6.25 billion billion) electrons per second, is defined as one ampere of current. Since electrons travel at the speed of light (186,000 miles per second), there is essentially an instantaneous movement of current in a circuit.

Electrical current is analogous to the water current that is measured as the number of gallons of water flowing past any point in the system in one minute and requires a closed loop to operate.

- In an electrical circuit, current flow is produced by connecting a load (component) having a specific resistance to a source of voltage, thereby creating a *closed circuit.*
- Electrons per second (amperes) of electrical current flow is analogous to gallons per minute (gpm) of water flow.
- Just as a *flow meter* can measure water current, an *ammeter* can measure electrical current. (See Figure 3.11)

When the value of the source voltage and the resistance of the load are known, the current flowing in a circuit can be calculated with the use of Ohm's Law. (See Chapter 4 - Ohm's Law)

The term that was originally used to identify current flow was *current intensity* and was designated with the letter **I**. Although "intensity" is no longer used, the letter "**I**" still remains the conventional designation for electrical current flow.

Just as water current has a direction as it flows through a hose or pipe, electrical current has a direction as it flows through a circuit.

Current (I) is indicated by an arrow (⟶) pointing in the direction of current flow. Using the conventional method of indicating current flow, it is shown as starting from the positive terminal of the supply voltage, going through the load, and then flowing back to the negative terminal of the voltage supply. (See Figure 3.8)

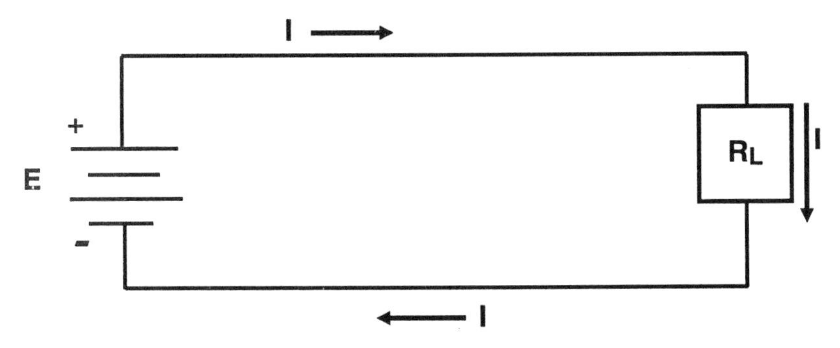

Current Designation and Direction
Figure 3.8

CONDUCTORS

An electrical **conductor** is a material that has the ability to electrically connect and carry current between two points in a circuit. This characteristic, called *conductance,* depends on the material and cross sectional area of the conductor. The larger the cross section of a conductor, the greater its conductance.

For all practical purposes, a conductor has essentially zero resistance and is shown graphically as a straight line. The resistance of a conductor is so small that it is considered to be negligible in comparison to the resistance of the load to which it is connected.

All references to a conductor in this text imply that the resistance of the conductor is considered to be zero, regardless of its length.

Conductors are used to electrically connect a voltage source, **E**, to a load. A closed circuit is created allowing current, **I**, to flow in a closed loop. (See Figure 3.9)

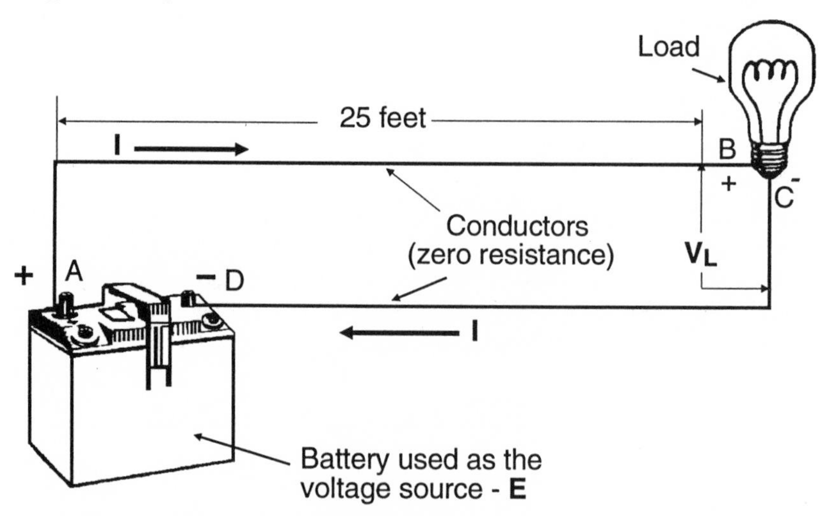

Connecting a Voltage Source to a Load to Create a Circuit
Figure 3.9

Since a conductor has essentially zero resistance, points **A** and **B** in Figure 3.9 are considered to be at the same point electrically as are points **C** and **D**. Therefore, whatever voltage, **E**, exists across the source terminals, the same voltage also exists across the load (light bulb) despite the fact that the voltage source and the load are physically far apart. The load voltage is designated as $\mathbf{V_L}$.

CONDUCTOR MATERIALS

Each conductor material has unique features that dictate its selection. Most metals have very low electrical resistivity (high conductivity) and are generally considered to be efficient conductors. For example, in critical applications where a conductor is exposed to a corrosive atmosphere, gold is generally used because of its non-corrosive, non-tarnishable qualities.

Copper is the most common material used for conductors. It has extremely low resistivity, is easily shaped, can be bent many times without breaking, and is easily soldered. Copper is relatively easy to process and is much less expensive than most other conductors.

Silver, having lower resistivity than copper, is used in applications requiring extremely low resistance, however, silver is susceptible to oxidation and can introduce undesired increased resistance at contact points. Silver is also more difficult to solder and is considerably more expensive than most other conductive materials with the exception of gold.

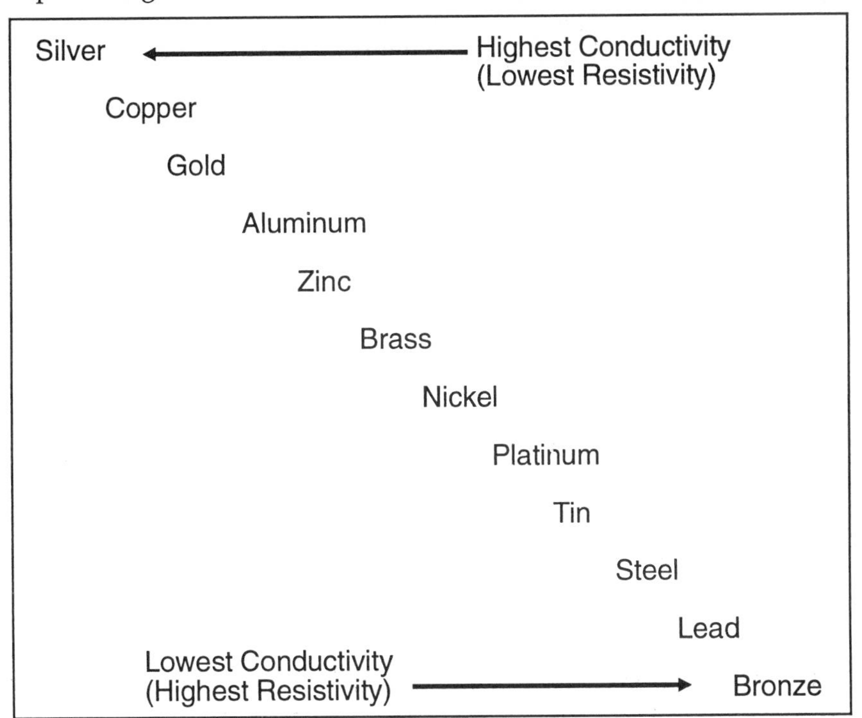

Conductor Materials in Order of Decreasing Conductivity
Figure 3.10

METERS

Voltage across a supply, a component, or across any two points of a circuit can be measured with the appropriate use of a voltmeter (either DC or AC). (See Figure 3.11)

The *voltmeter* (AC or DC), the *ohmmeter*, and the *ammeter* (AC or DC) are the three basic individual meters available for appropriate electrical measurement of voltage, resistance, and current.

These measurements can be combined in a single meter, referred to as a *multimeter*, providing all three measurement functions in one enclosure using only one meter display.

A variation of the multimeter is a *volt-ohmmeter* (VOM), designed to measure either AC or DC voltage and resistance.

Multimeters usually have a switchable control to select the desired measurement mode; some multimeters come with separate jacks on the meter panel for selecting the desired mode.

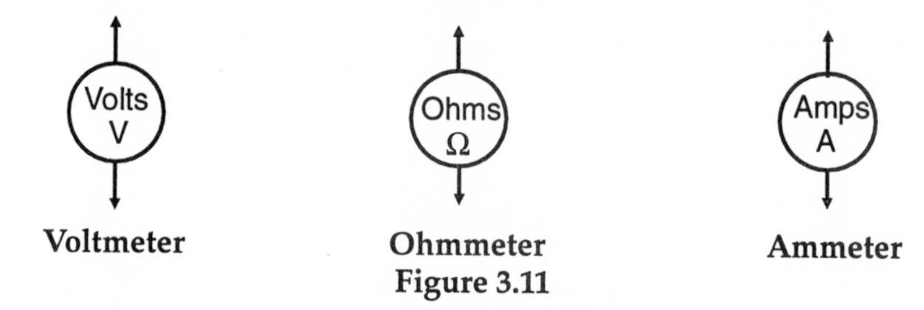

Figure 3.11

The use of a voltmeter and ammeter indicating the proper method of measuring voltage and current is shown in Figure 3.12.

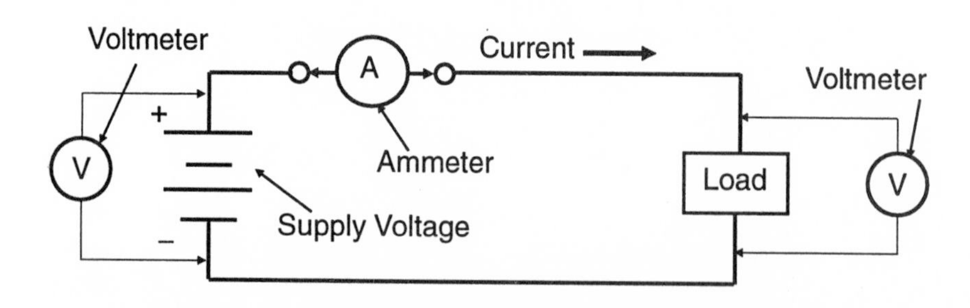

Illustrating the Use of a Voltmeter and Ammeter
Figure 3.12

GRAPHIC SYMBOLS

The English language uses graphic symbols, called letters, to represent its alphabet. These symbols, or letters, are arranged to form words and sentences for written communication. In a similar manner, the electronics language uses graphic symbols to represent its alphabet. The graphic symbols are used in a *circuit diagram* to represent the elements of a circuit or a *system*.

- A circuit diagram is also called a *schematic diagram* or *drawing*.

- A system is a configuration of any number of circuits, properly interconnected in a complete electrical or electronic structure, to perform one or more complex functions

SUBSCRIPT AND SUPERSCRIPT NOTATIONS

A **subscript** notation is a letter, a number, or a symbol written to the right and slightly below a letter, number, or symbol to indicate a specific component or circuit parameter. For example:

- $\mathbf{R_L}$ is the resistance (**R**) of the load (**L**).

- $\mathbf{V_C}$ is the voltage **V** across the capacitor (**C**).

- $\mathbf{C_2}$ is the **second** capacitor(**C**) in the designation of circuit parts on a schematic drawing, however, the designation of circuit parts can also be shown with no subscript as **C1**, **C2**, etc.

A **superscript** notation is a letter, a number, or a symbol written slightly above and to the right of a letter, number, or symbol to indicate a specific identification or numerical value. For example:

- $\mathbf{V^2}$ is the voltage (**V**) **squared** (the voltage raised to an exponential power of **2**), or that value of voltage multipled by itself: **V x V**.

- $\mathbf{2^4}$ is the number **2** used **4** times as the multiplication factor: 2 x 2 x 2 x 2 = 16.

- $\mathbf{10^n}$ is the number **10** used **n** times as the multiplication factor, where **n** represents any number.

4

CHAPTER
FOUR

OHM'S LAW

CIRCUIT RELATIONSHIPS

OHM'S LAW CALCULATIONS

CIRCUIT PROTECTION

GUARDING AGAINST:

- EXCESS CURRENT
- EXCESS VOLTAGE
- ELECTROSTATIC DISCHARGE (ESD)
- INSULATORS (NONCONDUCTORS)

OHM'S LAW

CIRCUIT RELATIONSHIPS

All electrical and electronic circuit design is based on the relationship between voltage, current, and resistance established by **Ohm's Law**. Ohm's Law provides a method to determine the value of an unknown parameter in a circuit.

Ohm's Law states:

The amount of electrical current, I, flowing in a circuit is equal to the electromotive force, or voltage, E, applied to that circuit divided by the total resistance, R, of the circuit. Ohm's Law is expressed by the following equation:

$$\textbf{Current} = \frac{\textbf{Voltage}}{\textbf{Resistance}} \quad \text{or} \quad I = \frac{E}{R}$$

The most elementary circuit, shown in Figure 4.1, is referred to as the *basic Ohm's Law circuit*. It consists of a supply voltage **E** (in volts), two conductors connecting the voltage to the load **R** (in ohms), resulting in a current **I** (in amperes).

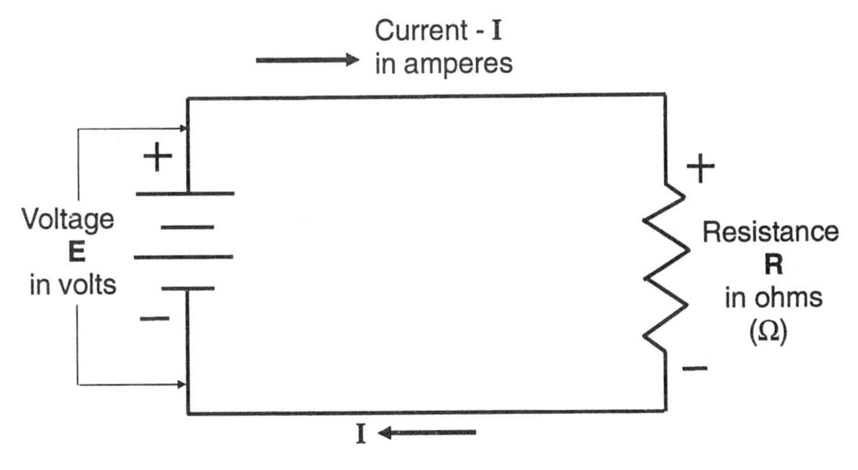

The Basic Ohm's Law Circuit
Figure 4.1

OHM'S LAW CALCULATIONS

In a circuit, the direction of current flow (**I**) is determined by the polarity (positive and negative terminals) of the supply voltage (**E**). Current flows from the positive (+) terminal, continues through the circuit load (**R**), and then reaches the negative (-) terminal of the supply voltage. The current (**I**) continues through the voltage source (**E**) to reach the starting point of the circuit loop.

In Figures 4.2, the current flows from point **A** (the plus terminal of the supply voltage) through conductor, **A-B**, and through the load, **R**. It then flows through conductor, **C-D**, to point **D**, the negative terminal of the supply voltage, and then through the supply voltage, **E**, to end up at the starting point, **A**.

Current continues to flow through the circuit in this manner as long as the voltage source is sustained and the circuit is closed, or complete. Since current travels at the speed of light, this process occurs instantaneously.

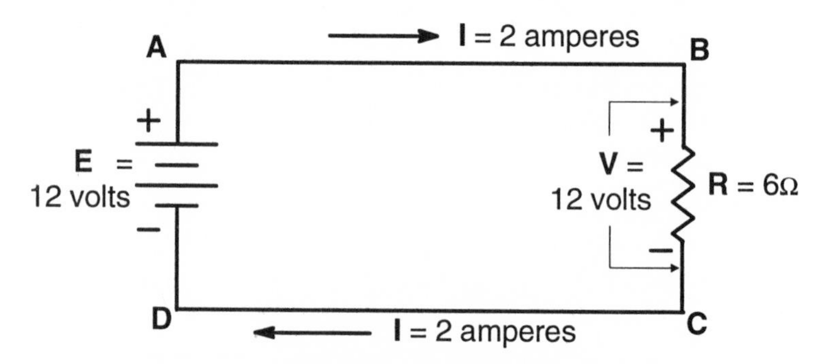

The Basic Ohm's Law Circuit
Figure 4.2

The first version of Ohm's Lawstates that:

$$\textbf{Current} = \frac{\textbf{Voltage}}{\textbf{Resistance}} \quad \textbf{or} \quad \textbf{I} = \frac{\textbf{E}}{\textbf{R}} = \frac{\textbf{V}}{\textbf{R}}$$

In the example of Figure 4.2, **E** = **V** = 12 volts and **R** = 6 ohms.

Using the Ohm's Law relationship:

current (**I**) is calculated as: $\frac{12}{6} = 2$ amperes

As the current (**I**), goes through the load (**R**), the current produces a voltage (**V**), across that load. In the basic Ohm's Law circuit of Figure 4.1, the load voltage (**V**), is equal to the supply voltage (**E**). If the supply voltage is unknown, this voltage (**E**) and the load voltage (**V**) can be calculated as follows:

Voltage = Current x Resistance or $E = V = I \times R$

This relationship is the second version of Ohm's Law.

If the resistance of a load (**R**) is unknown, its value can be calculated as the voltage (**V**) across the load, divided by the current (**I**) flowing through the load and is expressed as follows:

$$\text{Resistance} = \frac{\textbf{Voltage}}{\textbf{Current}} \quad \text{or} \quad R = \frac{E}{I} = \frac{V}{I}$$

This relationship is the third version of Ohm's Law.

These equations express the same Ohm's Law relationship in three different ways. The specific equation that is chosen is the one that will determine the unknown value when the other two values are known.

The relationships stated in Ohm's Law are the basis for all circuitry, ranging from a simple single-load circuit to an extremely complex computer circuit on a sophisticated, monolithic integrated circuit (IC) chip.

An easy way to remember the Ohm's Law relationship is to use the layout of the triangle shown in Figure 4.3.

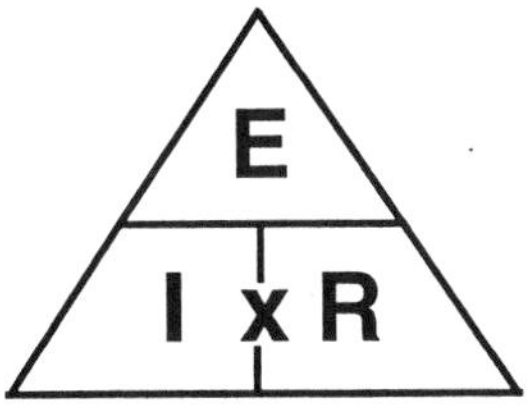

The Ohm's Law Relationship
Figure 4.3

With two values known, the unknown value is calculated by using the appropriate equation. The procedure becomes clear when the unknown parameter in the triangle is covered, indicating the applicable arithmetic process.

By covering **E**, the voltage is calculated as: **I** x **R**

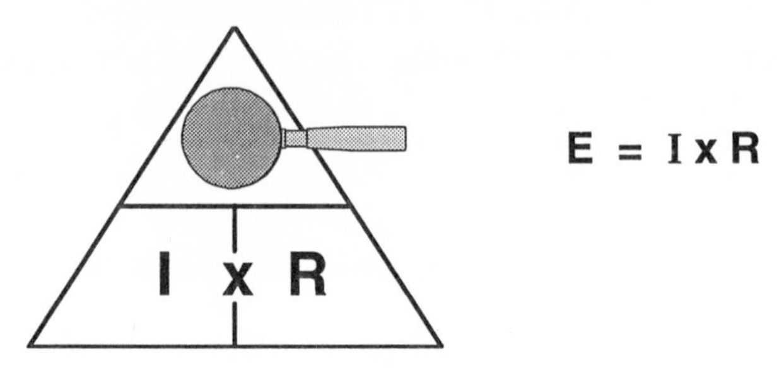

By covering **I**, the current is calculated as: $\frac{E}{R}$

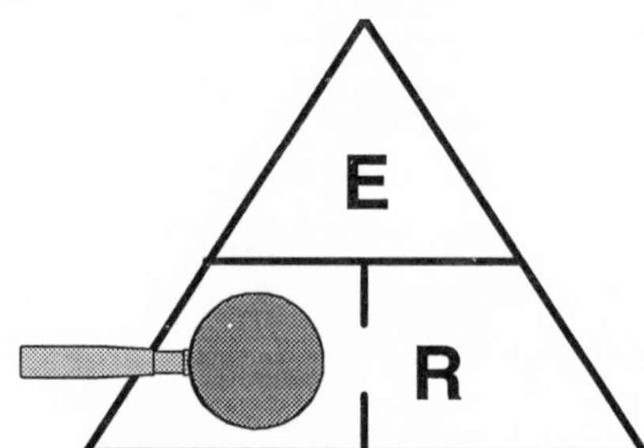

$$I = \frac{E}{R}$$

By covering **R**, the resistance is calculated as: $\frac{E}{I}$

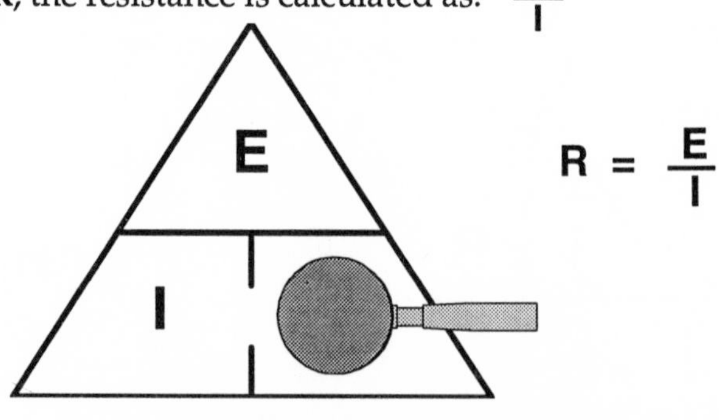

Figure 4.4

CIRCUIT PROTECTION

GUARDING AGAINST EXCESS CURRENT

FUSES AND CIRCUIT BREAKERS

A condition can arise in the basic Ohm's Law circuit (Figure 4.2) which causes both conductors to come in contact with each other.

> **When two conductors connected directly to a supply voltage touch each other, a condition is produced that bypasses the load. The original load is no longer functioning as the circuit load, but is replaced by a zero resistance path. This is called a *short circuit* or a "*short*" across the supply voltage.**

A short can also be caused by some malfunction of the load which reduces the load resistance to zero. Since the resistance across the voltage supply is now zero, the current is no longer limited and begins to rise suddenly toward infinity. If the resistance, **R**, is equal to **zero**, the current, **I**, is equal to **infinity**.

If any number is divided by zero, the result is infinity, symbolized by the symbol ∞. By definition, infinity is a hypothetical, limitless, and immeasureable number or quantity.

$$I = \frac{E}{R} = \frac{E}{0} = \infty$$

A conductor has a rated maximum current capability that should not be exceeded. If the current increases toward infinity because of a short circuit, one or both of the conductors in the circuit might melt. In this condition, the extremely high current may produce excessive heat that might cause a fire, resulting in danger to life and property. To avoid a current overload condition, the circuit must be protected.

Even if the conductors are assumed to be infinite in diameter (with a theoretically infinite current capability), the supply voltage would not be able to sustain an excessively high current demand or overload current. The circuit would stop functioning and, possibly, become permanently damaged. In addition, an excessively high current condition could severely damage or burn out any element of the circuit.

A **fuse** or a **circuit breaker** is an effective safeguard against a short circuit when installed in series with one of the conductors. If the rated current value of either one is exceeded, the fuse will melt, or the circuit breaker will release, thereby opening the circuit. The current will be reduced to zero, preventing damage to the conductors, the voltage source, and other circuit components.

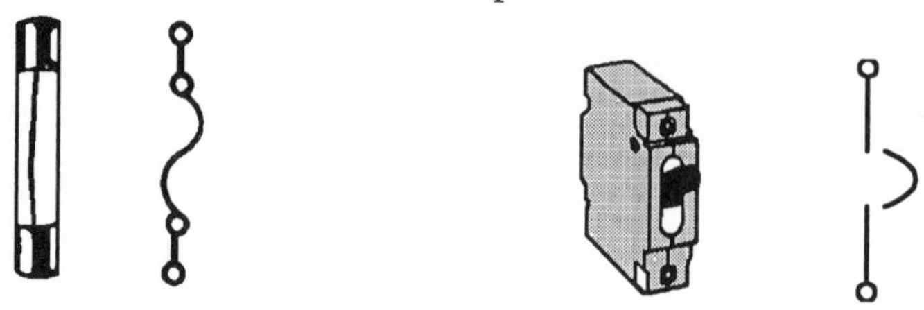

Fuse and Symbol **Circuit Breaker and Symbol**

Figure 4.5

The criteria for selecting the maximum operating current for a conductor is determined by its material, diameter, and the environment in which it will be used. In all cases, the maximum current capacity of a conductor must not be exceeded. Standards for the safe limits of current capacity for conductors are determined by the Underwriter's Laboratory (UL) and Canadian Standards Agency (CSA).

The rating for a fuse or circuit breaker is selected to provide circuit protection by limiting the current in that circuit to a value at, or below, the maximum current capacity of the circuit components or the associated conductors, whichever is lower. If the maximum current capability of conductors is 15 amperes and the maximum current capability of the component(s) in a circuit is 20 amperes, a proper rating for a fuse or circuit breaker is 15 amperes. In this case, the protective device protects the circuit against current levels above 15 amperes through both conductor and components. If another load is added to the circuit, increasing the total current to more than 15 amperes, the protective device will trigger to protect against excessive conductor current.

The choice of either a fuse or a circuit breaker is generally determined by circuit design requirements, cost, convenience, and safety considerations.

- A fuse is initially less expensive, will "fail safe" by opening the circuit, but has to be replaced when melted.
- A circuit breaker initially costs more, sometimes will not "fail safe", but can be re-set to return the circuit to its normal operating condition without having to be replaced.

THERMISTOR OR POLYSWITCH

A *thermistor* or *polyswitch*, a normally low-resistance thermal resistor, may be used as an automatically resettable fuse to protect the circuit from excessively high surge (initial turn-on), or in-rush current. Surge current is often produced when a system is turned on, and if excessive, could possibly damage the circuit.

When a system is turned on, the surge current creates an increase in temperature in the thermistor or polyswitch that has a positive temperature coefficient. The increased temperature causes its resistance to increase at a rapid rate, which, in turn, reduces the circuit current to a safe value until the surge is no longer present. With decreased current, less heat is generated at the thermistor or polyswitch and its resistance is reduced to its normal operating value.

BIMETALLIC SWITCH

A *bimetallic switch* can act as a self-resetting circuit breaker to protect circuitry. Excessive current can cause an increase in heat across a bimetallic switch. With each metal section of the switch having a different coefficient of expansion, the two contacts of the switch will separate, momentarily opening the circuit. With no current flowing, the switch will cool, forcing the contacts together again to close the circuit. If the condition that caused the excessive current to flow no longer exists, the switch will stay in its normally closed state, and the circuit will function properly again.

GUARDING AGAINST EXCESS VOLTAGE

Metal oxide varistors and *zener diodes* are other devices that provide circuit protection against excessively high voltage, such as sudden bursts of high voltage in a power line (voltage transients or spikes) or lightning. Any voltage in excess of the rated voltage of the protective device will be effectively shorted to ground, or absorbed to provide circuit protection.

These devices, particularly the zener diode, have extremely fast response time, coupled with the ability to absorb huge amounts of electrical energy for a short period of time. The subject of zener diodes is discussed in greater detail in Volume Two - Part I - Discrete Semiconductors.

GUARDING AGAINST ELECTROSTATIC DISCHARGE (ESD)

When certain dissimilar insulating materials are rubbed together, friction is produced between them, causing electrons from one material to be transferred to the other. This accumulation of electrons on the surface of a material is due to a phenomenon called the *triboelectric effect*, referred to more commonly as an *electrostatic charge*.

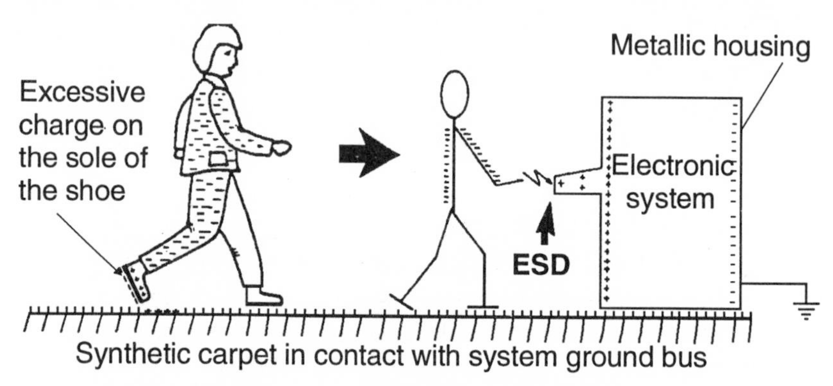

Triboelectric Effect and Generating Electrostatic Discharge (ESD)
Figure 4.6

Depending on the triboelectric tendency of the materials involved, different quantities of electrons will be transferred between materials. Typical triboelectric materials include: glass, asbestos, mica, wool, silk, paper, cotton, rubber, and a variety of plastics.

If the material with an excess of electrons comes in physical contact with a material with fewer electrons, the electrons in the material with the greater number of electrons will discharge into the material having a fewer number of electrons. A neutralization, or balance, of the built-up charge will occur.

The transference of electrons from one insulating material to another, either conductive or insulating, is called *electrostatic discharge* (ESD); its energy takes the form of voltage and current. The current, generally, is very low but the electrostatic voltage could be quite high.

Under certain dry atmospheric conditions, this voltage could be higher than 30,000 volts. Electrostatic discharge at levels of 30 volts and higher can be disastrous to electronic equipment, particularly computers. Typical problems caused by ESD include errors in keyboard data entry, changing of memory data, incorrect instructions and, sometimes, damage to a complete system. Any electronic circuit, however, can be completely disrupted or shut down by electrostatic discharge when it has caused failure of sensitive circuit components in the system.

Several factors contribute to the damaging effects of ESD. Reduced humidity creates the most severe results. At a relative humidity of 65% or higher, there is very little electrostatic discharge. If the relative humidity in the area is below this critical value, the effects of ESD could be detrimental to circuitry. At 20% relative humidity, 300 times more electrostatic energy is generated than at 65% relative humidity.

When rubbing nylon and polyester material, static voltages as high as 6500 volts can generated.

A variety of approaches used to protect against damage from ESD center mainly around different techniques of dissipating or neutralizing this energy. Carpets and rugs are made with special antistatic materials and coatings to "bleed" away static buildup. Plastic materials are sprayed with antistatic solutions. Metal wrist straps connected to ground are worn by assembly operators in some semiconductor factories and antistatic tables and trays are used.

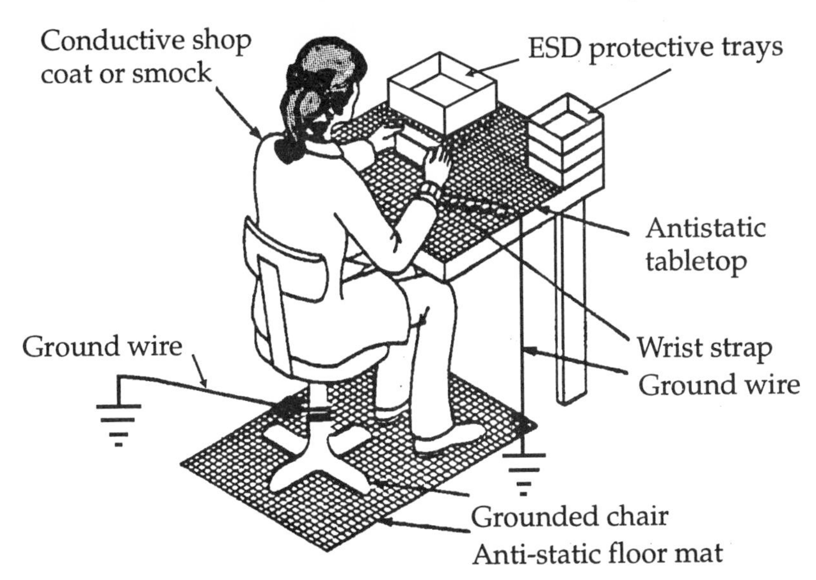

Minimizing Electrostatic Discharge at a Workstation
Figure 4.7

MOSFET semiconductors are particularly sensitive to the effects of ESD. Antistatic trays are used to carry these components within the assembly and test areas. MOSFET devices are shipped in special antistatic material with low triboelectric propensity.

A very effective technique to protect against ESD is to maintain the relative humidity in a work area above the critical value of 65%. All protective techniques are useful; special zener diodes with extremely fast response to an overvoltage condition are particularly useful as protective devices against electrostatic discharge. These zener diodes act extremely fast to short out electrostatic voltages higher than their rated voltage. They have sufficient power capability to absorb the short-interval energy generated by the offending electrostatic discharge. Above their rated voltage, these zener diodes act as a short across the circuit to which they are connected, but have no effect on that circuit below the rated voltage.

The characteristics, specifications, and other protective techniques for MOSFETs and zener diodes are discussed in Volume Two - Part One - Discrete Semiconductors.

INSULATORS (NONCONDUCTORS)

Placing insulating material around a bare conductor, or between conductive elements of a circuit, will protect the circuit and prevent a possible short circuit from occurring. *An insulator is called a nonconductor and is defined as a material that has essentially infinite resistance.* Its resistance depends on its molecular structure and dimensions. Many types of nonconductors are used in electrical and electronic circuitry with the commonly used types listed below.

Ceramic	Air	Aluminum Oxide
Rubber	Polystyrene	Tantalum Oxide
Mica	Nylon	Beryllium Oxide
Teflon	Quartz	Polypropylene
Neoprene	Mylar	Glass (Silicon Dioxide)
Enamel	Paper	Polyvinyl Chloride (PVC)

If two insulated conductors come in contact with each other at the insulated sections, the circuit is not affected in any way. No short circuit occurs since the insulation is considered to have infinite resistance, and the circuit remains undisturbed.

In applying Ohm's Law, the voltage **(E or V)** is divided by an infinite resistance, **R**, resulting in zero current flow.

Since $\mathbf{I} = \dfrac{\mathbf{E}}{\mathbf{R}}$ and **R** = **infinity,** then **I** = **zero.**

If the insulating materials (nonconductors) have been damaged, or in some way removed from the conductors and/or components that were previously protected, a short circuit will then occur if the components, or previously protected conductors, are allowed to touch each other. Components and conductors can be permanently damaged and circuit operation adversely affected if proper circuit protection has not been provided.

5

CHAPTER
FIVE

CIRCUIT CONFIGURATIONS

SERIES CIRCUITS

PARALLEL CIRCUITS

SERIES/PARALLEL CIRCUITS

CIRCUIT CONFIGURATIONS

An electrical or electronic circuit consists of a single load, or many loads, connected to a voltage source. Current will flow from the voltage source, through the load, and back to the voltage source when the circuit is operating (closed circuit).

In the basic Ohm's Law circuit (See Chapter Four, Figure 4.1), a single load is connected to a supply voltage in a closed loop. When additional loads are added to this circuit, the manner in which they are connected is called the *circuit configuration.*

With additional loads, a more functional circuit can created and configured in two different ways. Depending on how the loads are connected, the resulting configuration can produce either a **series** or **parallel** circuit. Any combination of the two in a single circuit is called a **series/parallel** circuit.

To analyze any circuit it is necessary to determine the values of load current, load voltages, and other significant parameters of the circuit. Specific rules and relationships are applied for each configuration. Knowing these values, the proper components and conductors can be selected for optimized circuit design.

Simple or complex circuits are usually drawn in schematic form to be read from left to right. There is, however, no fixed rule as to where the symbols for either the voltage source or the circuit components should be placed on the drawing.

SERIES CIRCUITS

> If two or more loads are connected to a source of voltage in a chain-like manner, with one load following the other, the resulting configuration is called a **series circuit.**

Since a series circuit has more than one load, to calculate its load current it must be changed to a simpler form called its *equivalent circuit*. The equivalent circuit is drawn in the form of the basic Ohm's Law circuit. To create and analyze an equivalent circuit, the values of the supply voltage, **E**, and the load resistances, $\mathbf{R_1}$ and $\mathbf{R_2}$, must be known. (See Figure 5.1)

Rule One of a Series Circuit:
The equivalent load resistance (R_E) in a series circuit is the sum of the resistances of the individual loads: $R_E = R_1 + R_2 + ..$ etc.

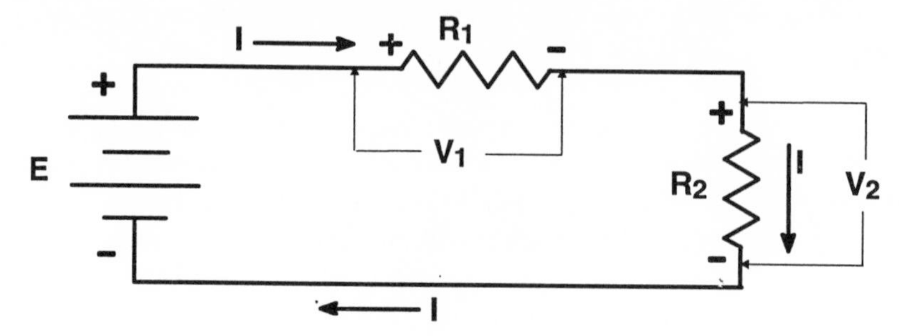

Two-load Series Circuit
Figure 5.1

Once the equivalent resistance value is calculated, the basic Ohm's Law circuit can be drawn. (See Figure 5.2) The current (I) is calculated by applying the Ohm's Law relationship, **I = E/R.**

The calculated load current (I) in the equivalent circuit is the same as the load current (I) coming from the voltage supply (E) in the original series circuit.

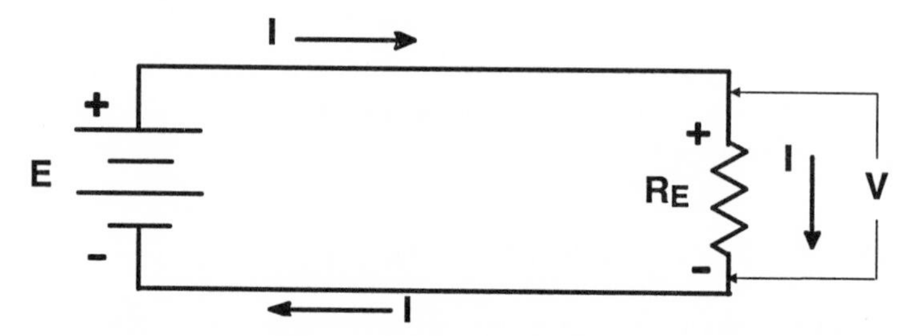

Equivalent Circuit of the Two-load Series Circuit of Figure 5.1
Figure 5.2

The path of the current flow in Figure 5.1 is as follows:

- The current (I) leaves the positive terminal of E, goes through $\mathbf{R_1}$, to produce a voltage (sometimes called a "voltage drop") across R_1. This voltage is equal to $\mathbf{I \times R_1}$ and designated as $\mathbf{V_1}$.
- The current (I) continues through $\mathbf{R_2}$ and produces a voltage across $\mathbf{R_2}$ (the product of $\mathbf{I \times R_2}$) and is designated as $\mathbf{V_2}$. The polarities of the voltages, $\mathbf{V_1}$ and $\mathbf{V_2}$, are shown in Figure 5.1.

- In a series circuit, the current from the voltage source will remain the same as it flows through the circuit and will produce a voltage across each load. The value of each load voltage is the product of the current through that load times its resistance.

$$V_1 = I \times R_1 \; ; \; V_2 = I \times R_2 \; ; \; ..... \text{ etc.}$$

Rule Two of a Series Circuit:

The voltage supply (or source voltage) in a series circuit is equal to the sum of the voltages across each individual load.

$$E = V_1 + V_2 + ... \text{ etc.}$$

The total of the individual load voltages must equal the value of the supply voltage.

SERIES CIRCUIT EXAMPLE

In the series circuit of Figure 5.3, Ohm's Law is applied to calculate the current in the circuit and the voltages across each load.

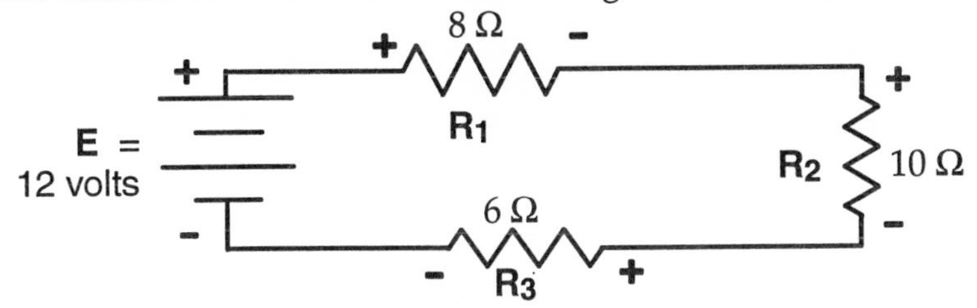

Three-load Series Circuit Example
Figure 5.3

A voltage supply of 12 volts is connected to three loads (**R_1**, **R_2**, and **R_3**) in series. The value of these loads are 8, 10, and 6 ohms, respectively. The equivalent resistance is equal to the sum of the resistances of the three loads; in this case, 24 ohms. In the equivalent circuit of Figure 5.4, the current, **$I = E/R$** = 24/12 = ½ (0.5) ampere.

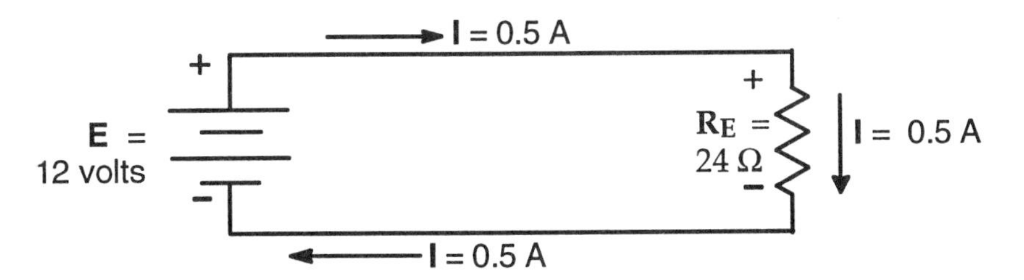

Equivalent Circuit of the Circuit of Figure 5.3
Figure 5.4

The circuit of Figure 5.3 is redrawn in Figure 5.5 with the calculated current, I= 0.5 amperes, added to the drawing.

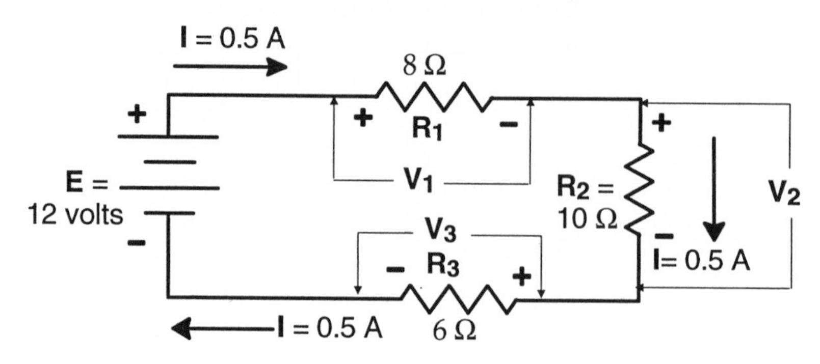

Calculations of Individual Load Voltages
Figure 5.5

The current (**I**) leaves the positive terminal of the supply and flows through each load, producing voltages of 4, 5, and 3 volts respectively. These voltage values are added, resulting in a sum of 12 volts, reaffirming the series circuit relationship:

$$\mathbf{E = V_1 + V_2 + V_3} = 4 + 5 + 3 = 12 \text{ volts}$$

If an additional load is inserted in series into the circuit, the circuit conditions will change. The extra load adds to the original equivalent resistance and increases it to a higher value. The additional resistance causes a reduction in load current, changing the voltage across each load in the new circuit.

If a load is removed from the original circuit, or shorted out, thereby reducing this particular load value to zero, the equivalent resistance of the circuit decreases, causing more current to be drawn from the supply. If one of the loads burns out or opens, the circuit is now an *open circuit*; no current flows and the circuit stops functioning.

PARALLEL CIRCUITS

If two or more loads are connected directly across a voltage supply, the configuration is called a **parallel circuit**.

Parallel circuits are more commonly used than series circuits and is the preferred configuration for the electrical circuits of appliances, accessories, and equipment in automobiles, homes, and offices. In parallel circuits, the power supply for all appliances is the same and all other other loads will still operate if any one burns out.

Rule One of parallel circuits (see Figure 5.6):
All voltages in parallel circuits are equal to each other in both polarity and magnitude (or amplitude).

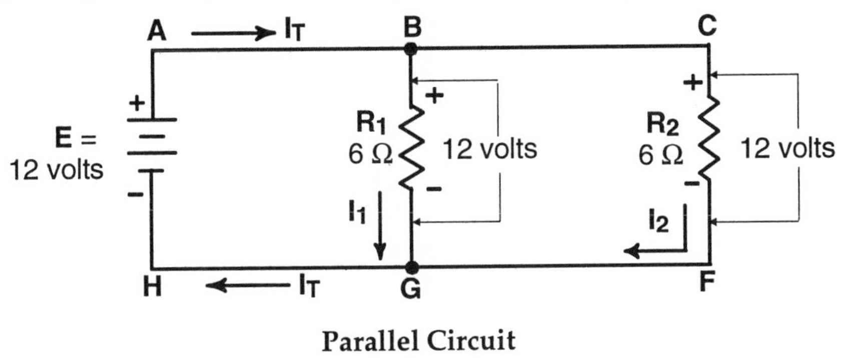

Parallel Circuit
Figure 5.6

Rule Two of parallel circuits:
In a parallel circuit, the total current (I_T) coming from the voltage source is equal to the sum of the individual currents flowing through each load.

$$I_T = I_1 + I_2 + \ldots\ \text{etc.}$$

The values of the supply voltage, **E** (12 volts) and R_1 (6 ohms) are given. The current, I_1, flowing through R_1 is calculated by the Ohm's Law equation:

$$I_1 = \frac{E}{R_1} = \frac{12}{6} = 2 \text{ amperes}$$

When a second load, R_2, is connected in parallel with (directly across) R_1, the voltage across R_2 will also be 12 volts since, in the circuit of Figure 5.6, points **A**, **B**, and **C** are at the same point electrically and points **F**, **G**, and **H** are at the same point electrically.

The voltages, V_1, across R_1, and, V_2, across R_2, are, therefore, equal to each other. Any other loads connected in parallel in this circuit will have the same voltage across them. This voltage is equal to the source voltage, **E**, since all the loads are in parallel with the supply.

$$E = V_1 = V_2 = \ldots\ \text{etc.}$$

Since parallel voltages are equal, the voltage, V_2, across R_2 (6 ohms) is also 12 volts and the current (I_2) through R_2, is also 2 amperes.

$\mathbf{I_2}$ is the additional current being supplied by the voltage source. If the source is supplying 2 amperes for $\mathbf{R_1}$ and 2 amperes for $\mathbf{R_2}$, the voltage source (**E**) must supply the sum of the currents in both loads, or the total current ($\mathbf{I_T}$) must equal 4 amperes to satisfy both loads.

As the total current ($\mathbf{I_T}$) leaves the positive terminal of the voltage source (**E**) it reaches the junction at point **B**. Here the current splits, with each individual current taking a different path; current $\mathbf{I_1}$ goes through $\mathbf{R_1}$ and current $\mathbf{I_2}$ goes through $\mathbf{R_2}$.

The individual current in each parallel path is equal to the voltage across each load divided by its resistance.

At point **G**, the two currents ($\mathbf{I_1}$ and $\mathbf{I_2}$) rejoin to equal the total current ($\mathbf{I_T}$). This current returns to the negative terminal of the voltage source, moving through the voltage source to its positive terminal where it began, completing the circuit.

Knowing the value of the supply voltage (**E**) and the total current ($\mathbf{I_T}$), the equivalent resistance ($\mathbf{R_E}$) of the parallel circuit can be calculated by using Ohm's Law.

$$\mathbf{R_E} = \frac{\mathbf{E}}{\mathbf{I_T}} = \frac{12}{4} = 3 \text{ ohms}$$

EFFECTS OF PARALLEL LOADING

- As more loads are connected in parallel, each new load will require current from the supply voltage, **E**. The total current ($\mathbf{I_T}$) will increase as the value of the equivalent circuit resistance ($\mathbf{R_E}$) is reduced.

- A load with a value equal to infinity will have no loading effect on the parallel combination.

- Whenever two or more loads are connected in parallel, the value of $\mathbf{R_E}$ is always less than the value of any individual load.

- Added parallel loading can continue until a point is reached where the supply is unable to furnish enough current for the circuit. The voltage source is then considered to be *overloaded* and the circuit will either stop functioning or function at a reduced level of voltage.

- To prevent an overload condition, a fuse or a circuit breaker would normally be used to protect the voltage source. Its capacity is chosen to be equal to the maximum current rating of the voltage source, the circuit components, and the conductors.

EQUIVALENT RESISTANCE IN PARALLEL CIRCUITS

Depending upon the number of loads in a parallel circuit and the resistance value of each load, the following approaches may be used to calculate equivalent resistance of parallel circuits. For example:

- If all loads in parallel have the same resistance value (**R**), the equivalent resistance is equal to that same value divided by the number of loads, **n**.

$$R_E = \frac{R}{n}$$

In the circuit of Figure 5.6, each load resistance is 6 ohms. By using this equation, the equivalent resistance can be calculated as:

$$R_E = \frac{6}{2} = 3 \text{ ohms}$$

- If there are only two loads in parallel, each having a different value, the equivalent resistance can be calculated as:

$$R_E = \frac{R_1 \times R_2}{R_1 + R_2}$$

- If there are three or more loads in parallel and they are not all equal in value, the equivalent resistance can be calculated as:

$$R_E = \frac{1}{\frac{1}{R_1} + \frac{1}{R_2} + \frac{1}{R_3} + \ldots \text{etc.}}$$

This last equation is called the "universal" equation for the equivalent resistance of a parallel circuit. It can be used with two, three, or more loads in parallel, regardless of their value.

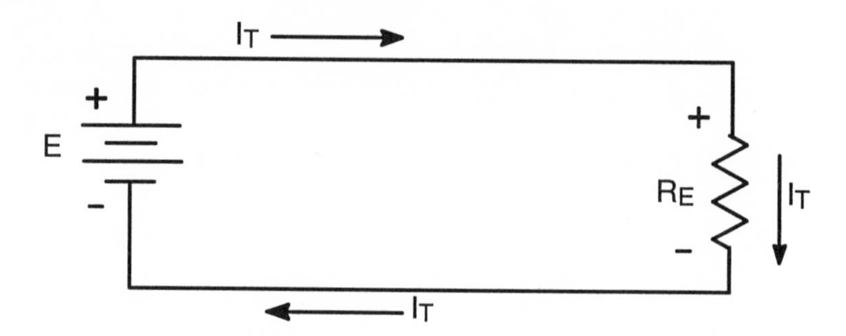

Equivalent Circuit of the Parallel Circuit of Figure 5.6
Figure 5.7

PARALLEL CIRCUIT EXAMPLE

The parallel circuit shown in Figure 5.8 is an example of a portion of an automobile's electrical system. The accessories, having known resistance values, are connected in parallel across a 12 volt battery.

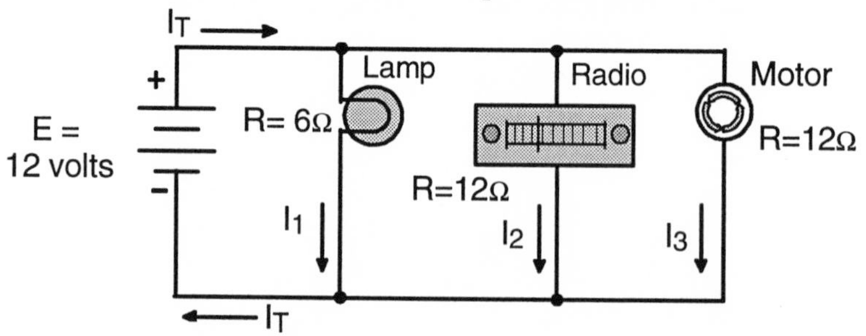

Partial Electrical System of an Automobile
Figure 5.8

The current through each load is calculated by Ohm's Law with no need to reduce the circuit to its equivalent form. Since all voltages in parallel are equal, 12 volts are across each accessory. The current is identified by the appropriate subscript and calculated as follows:

$$\mathbf{I_1} = \frac{12}{6} = 2\text{ A};\ \mathbf{I_2} = \frac{12}{12} = 1\text{A};\ \mathbf{I_3} = \frac{12}{12} = 1\text{A}$$

The total load current, $\mathbf{I_T}$, coming from the battery is equal to the sum of the currents in each load.

$$\mathbf{I_T} = \mathbf{I_1} + \mathbf{I_2} + \mathbf{I_3} = 2 + 1 + 1 = 4\text{ A}$$

With the total load current determined, the required conductor size, the required capacity of the battery, and the proper fuse size for protection of the system can now be selected. Since the total current is 4 amperes, a common, single fuse rated at 5 amperes may be used in the conducting path coming from the battery.

If one of the accessories develops a short circuit, the fuse will melt, opening the circuit and removing the voltages from all the loads.

An alternate design approach calls for the installation of a fuse in each accessory circuit, with each fuse having the proper rating for its own load.

For example: A 3 ampere fuse could be used for I_1, a 2 ampere fuse for I_2, and a 2 ampere fuse for I_3. (See Figure 5.9)

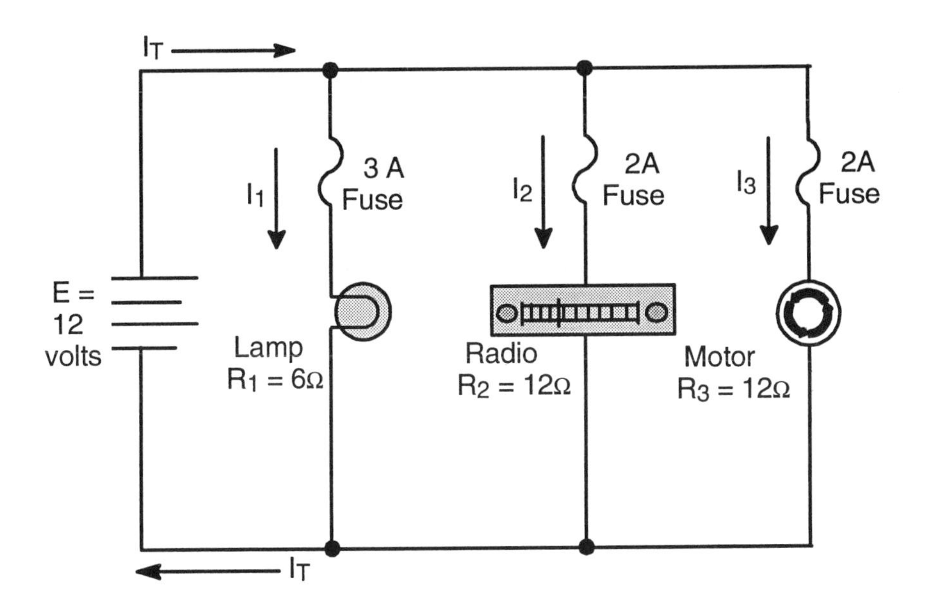

Portion of Automobile Electrical System
With Installation of Additional Fuses
Figure 5.9

In this arrangement, if one of the loads was shorted, only the fuse in that section of the circuit would burn out and the rest of the system would continue to operate.

SERIES/PARALLEL CIRCUITS

Most electronic systems are configured to provide a combination of both series and parallel circuits. Each section of the system is structured in this manner to perform a specific function within the system. This combined configuration is called a *series/parallel circuit* with each section following the rules of the corresponding series and parallel configurations.

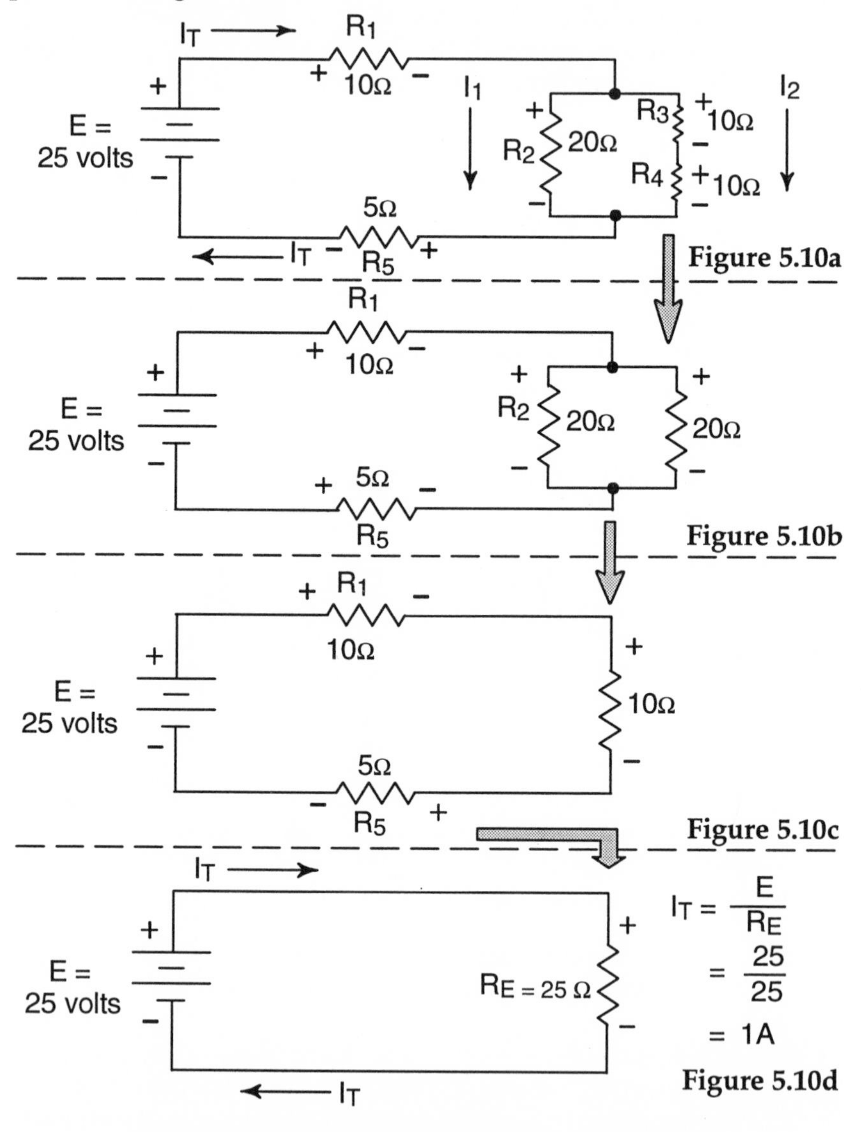

Series/Parallel Circuit and Equivalent Circuits
Figure 5.10

An example of this circuit and the process of simplifying it to finally create a basic Ohm's Law equivalent circuit is shown in Figure 5.10.

If the loads **R_2**, **R_3**, and **R_4** are treated as a single section, the circuit is seen to be a three-load series circuit, with **R_1** in series with the section **R_2**, **R_3**, and **R_4**. Load **R_5** is the last of the series loads.

The section **R_2**, **R_3**, and **R_4** consists of load **R_2** in parallel with the two loads, **R_3** and **R_4**, that are connected in series.

- The first equivalent of the original circuit (Figure 5.10a) is shown in Figure 5.10b. It is achieved by summing up the two 10 ohm series resistors, **R_3** and **R_4**, for an equivalent value of 20 ohms.
- The circuit in Figure 5.10b can be further simplified to create the equivalent circuit of Figure 5.10c by reducing the two 20 ohm parallel loads to their equivalent value of 10 ohms (R/n).
- The circuit in Figure 5.10c, consisting of three series loads, can now be reduced to the final equivalent circuit of Figure 5.10d by summing up the value of these three loads to equal 25 ohms.
- From this basic Ohm's Law circuit, the current **I_T** can be calculated by applying Ohm's Law:

$$I_T = \frac{E}{R} = \frac{25}{25} = 1 \text{ ampere.}$$

Going back to the original circuit shown in Figure 5.10a, the total current, **I_T**, in this circuit is the same as the **I_T** calculated in the final equivalent circuit shown in Figure 5.10d.

- **I_T** (1 ampere) leaves the positive terminal of the voltage source, E, and flows through **R_1**. Voltage, **V_1**, is produced across **R_1** which is equal to **I_T x R_1** (1 x 10) or 10 volts.
- **I_T** reaches the junction of **R_1**, **R_2**, and **R_3** and splits into two currents, **I_1** (through **R_2**) and **I_2** (through **R_3** and **R_4** in series).
- **I_1** and **I_2** rejoin at the junction of **R_2**, **R_4**, and **R_5** and **I_T** (1 ampere) flows through **R_5** to produce 5 volts across this load. The total current then returns to the negative terminal of the voltage source

to complete the circuit. The entire process occurs instantaneously, since current flows at the speed of light (186,000 miles per second).

According to Rule Two of series circuits, all the individual load voltages in the circuit must add up to equal the supply voltage, **E**, which, in this case, equals 25 volts.

- Since the voltage across $\mathbf{R_1}$ is equal to 10 volts, and the voltage across $\mathbf{R_5}$ is equal to 5 volts, the voltage across the section, $\mathbf{R_2}$, $\mathbf{R_3}$, and $\mathbf{R_4}$, must be 10 volts (25 volts - 15 volts). (See Figure 5.10a)

- With the voltage across $\mathbf{R_2}$ equal to 10 volts, the current $\mathbf{I_1}$ flowing through load $\mathbf{R_2}$ is calculated as $\mathbf{V_{R2}/R_2}$, or 10/20, which is equal to ½ ampere.

- The current, $\mathbf{I_2}$, flowing through loads $\mathbf{R_3}$ and $\mathbf{R_4}$ in series is also equal to ½ ampere since $\mathbf{I_2 = I_T - I_1}$.

- With the value of $\mathbf{I_2}$ now determined to be ½ ampere, the voltages across $\mathbf{R_3}$ and $\mathbf{R_4}$ can be calculated as 5 volts each by using Ohm's Law ($\mathbf{V = I \times R}$).

The circuit of Figure 5.10a is used an example of a series/parallel configuration. It illustrates how Ohm's Law is used to determine the values of the individual voltages and currents in the circuit and does not represent any specific application.

6 CHAPTER SIX

POWER

WORK AND POWER

ELECTRICAL POWER RELATIONSHIPS

HEAT

POWER

WORK AND POWER

In everyday language, the term "work" means any form of labor that produces a result. In a scientific context, work is defined as *the energy necessary to perform a specific task.*

In mechanics, work, or energy, is expressed as the force required to move an object for a specific distance. The terms "work" and "energy" are interchangeable and are expressed by the same unit of measurement - *foot-pounds.*

It takes 100 foot-pounds of work in the form of potential energy to raise a one pound weight to a height of 100 feet. When the weight is dropped, 100 foot-pounds of kinetic energy is generated to cause the weight to hit the ground. The duration of time in which the kinetic energy is used specifies the power that was available. The weight being moved is referred to as the mechanical load.

An example of mechanical power is shown in Figure 6.1 and is calculated as the product of force times distance per second.

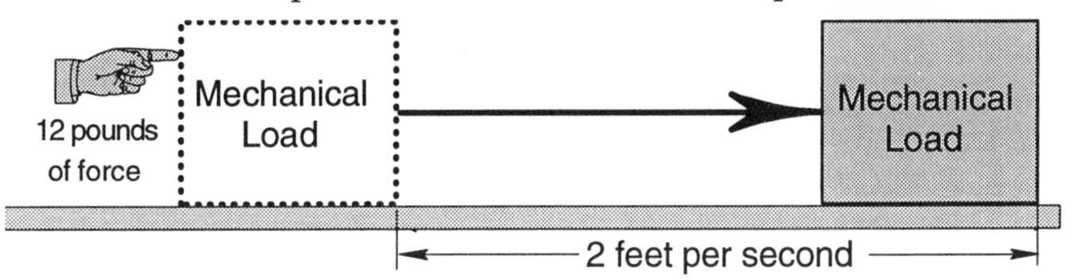

Illustration of Mechanical Power
Figure 6.1

When a mechanical force of 12 pounds is exerted on an object, moving it 2 feet in 1 second, the mechanical power generated is equal to the product of the force exerted on the object being moved times the distance it was moved in one second.

$$\text{Power generated} = \text{Power consumed}$$

$$12 \text{ pounds of force} \times 2 \text{ feet per second} = 24 \ \frac{\text{foot-pounds}}{\text{second}}$$

The mechanical power generated is equal to the mechanical power dissipated, consumed, or used by the system and is referred to as foot-pounds per second. This rarely used term can be converted into the more commonly used term of horsepower (hp). One horsepower is equal to 550 foot-pounds per second.

In electrical and electronic circuits, the values of the voltage and current required by the components must be determined to specify the specific components required. In addition, the power consumed by the components must also be determined for proper selection.

Power is defined as the amount of mechanical or electrical work performed (or the amount of energy expended) during a specified period of time.

ELECTRICAL POWER RELATIONSHIPS

Electrical power is similar to mechanical power since the electrical force of voltage, in volts, is analogous to mechanical force, in pounds. Current, in amperes (electrons per second), is analogous to the movement of the mechanical load in feet per second. The product of two electrical parameters in the circuit, voltage and current, produces electrical power, designated with the letter **P**.

In Figure 6.2, the supply voltage, **E**, is connected to a load, $\mathbf{R_L}$, to produce a load current, $\mathbf{I_L}$. The power generated, **P**, is equal to the power consumed and is calculated as the product of the two known electrical quantities, voltage and current.

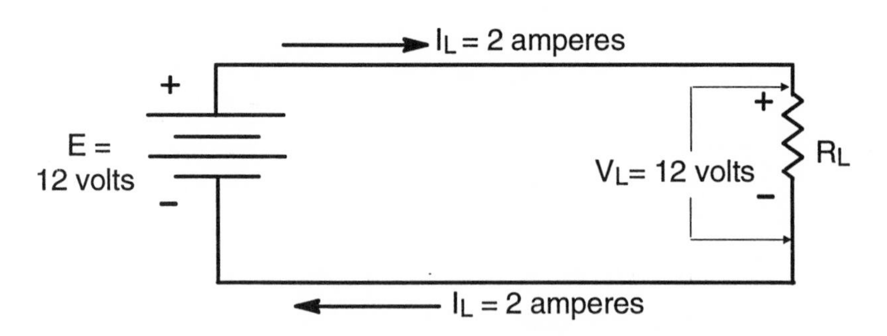

Illustration of Electrical Power
Figure 6.2

$$\mathbf{P = E \times I_L}$$

12 volts x 2 amperes = 24 volt-amperes = 24 watts

Instead of using volt-amperes as the dimension for electrical power, the term *watt*, named after **James Watt**, is more commonly used.

- Since **E** = **V** in a single-load or in a parallel circuit, the symbol for load voltage, **V**, can be substituted for the symbol, **E** , to determine the power consumed in any load, R_L.

- If the current, **I**, is unknown and values for load voltage, **V**, and load resistance, **R**, are known, power can be calculated by substituting the Ohm's Law equivalent for **I (V/R)** in the power relationship.

$$P = V \times I = V \times \frac{V}{R} = \frac{V^2}{R}$$

- The power relationship can also be expressed in terms of the current through a load and the resistance of that load. If the Ohm's Law equivalent of the voltage, **V**, **(I x R)** is substituted in the original power equation (**P** = **V** x **I**), then the power relationship becomes:

$$P = V \times I = (I \times R) \times I = I^2R$$

HEAT

When current flows in a component having an inherent resistance, **R**, heat is generated from the component and is calculated as I^2R. Unless there are specific requirements for the existence of heat in some part of a circuit, heat generally produces a destructive effect on the components in a circuit and affects circuit and system operation adversely.

Heat is the enemy of electrical and electronic circuits and systems.

The presence of heat in or near a component can change its characteristics and reduce its ability to handle its rated power. Heat can damage a component and other components in its vicinity and reduce or destroy the reliability of a system.

To maintain effective circuit operation and system reliability, the effects of heat must be minimized and effectively eliminated by the appropriate means of cooling.

To achieve maximum heat reduction, the following must be considered:

- Proper selection of components
- Appropriate circuit design
- Careful mechanical layout of a system
- Implementation of appropriate cooling techniques

Cooling techniques include:

- Mounting a component onto a flat metal plate, a finned assembly, or a metal chassis to transfer the heat generated in the component to some other area. This heat-transferring structure is referred to as a *heat sink*. When mounting a component on a heat sink, thermal paste is often used to ensure an effective transfer of heat between the component and the heat sink.
- Utilizing a fan or blower to move cool air over a component to remove or vent the accumulated heat.
- Installing the entire electronic system in an air-conditioned room to operate at a lowered temperature to affect the removal of heat.
- Installing a water-ducting system within the electronic system to provide a means of cooling.

The initial circuit design and system layout must take into account the problem of heat transfer. Regardless of which technique is used, the effects of heat must not be overlooked and treated as an afterthought at the completion of the design effort. This omission can prove to be costly and destructive to the system.

CHAPTER
SEVEN

ALTERNATING CURRENT AND DIRECT CURRENT

ALTERNATING CURRENT - AC

- THE REFERENCE
- THE SINE WAVE
- FREQUENCY, PERIOD, AND WAVELENGTH
- DEFINING ALTERNATING CURRENT (AC)
- THE FREQUENCY SPECTRUM

DIRECT CURRENT - DC

- DEFINING DIRECT CURRENT (DC)
- PULSATING DC VOLTAGE - HALF AND FULL WAVE
- STEADY-STATE DC VOLTAGE
- REGULATION
- PULSED DC VOLTAGE (DIGITAL PULSES)

ALTERNATING CURRENT AND DIRECT CURRENT

ALTERNATING CURRENT (AC)

THE REFERENCE

When investigating the principle of alternating current (AC), the concept of a *reference* must first be examined. For example, when a plane is flying at 35,000 feet, it is at a specific height, or *altitude,* with respect to sea level, its fixed reference. The concept of the reference is implied but is not usually stated. As the plane descends, its altitude changes with respect to the fixed reference, indicating a specific height above sea level.

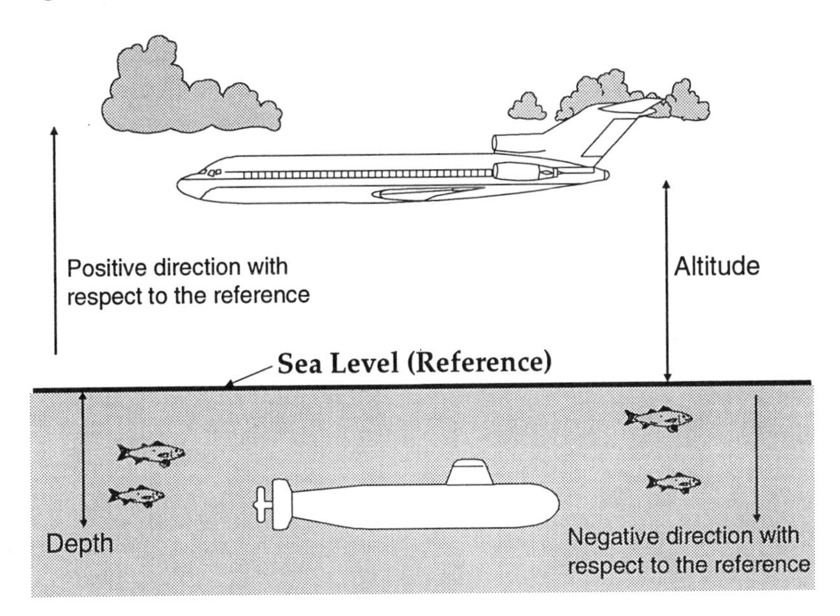

Figure 7.1

As another example, a submarine could be at a depth of 60 feet below sea level while a plane can be 60 feet above sea level. Although they are situated in opposite directions, they are both equidistant from the common reference.

Two factors must be considered in identifying the position of either object - the *distance* from a reference and the *direction,* or *location,* with respect to a reference. (See Figure 7.1)

In an electrical or electronic circuit or system, the direction of a voltage with respect to a reference is called its *polarity*. The amount of voltage with respect to a reference is called its *magnitude* (for DC voltages), or *amplitude* (for AC voltages). The term "ground" is used interchangeably with the reference in most systems.

- A voltage above a reference (ground) is called *plus* (*positive polarity*)
- A voltage below a reference is called *minus* (*negative polarity*).

Power companies throughout the world establish a reference for AC power by electrically connecting one of the two terminals of their generator to the earth (ground). The earth is composed of many elements and moisture; it is considered to be a low-resistance conductor that establishes a common reference for power distribution.

Since a voltage exists across two terminals of a generator, one terminal (the *ground* terminal) is connected to the earth with a large conductor and copper rod. The other terminal (the *hot* terminal) is at a specific high voltage with respect to ground. Both terminals are connected to the power distribution line. (See Figure 7.2)

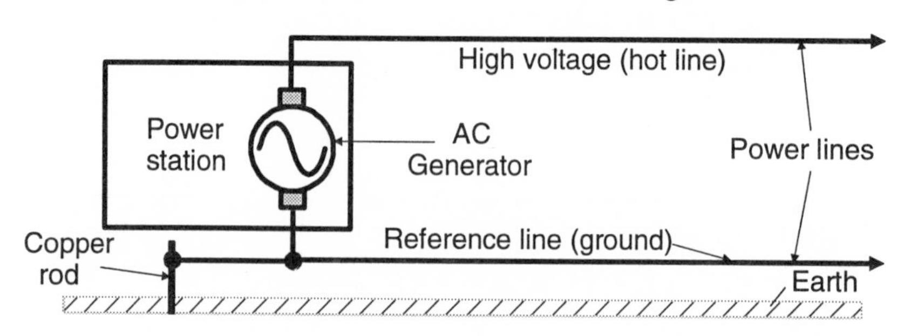

Simplified Power Distribution System
Figure 7.2

An automobile's steel chassis acts as the reference for its 12 volt electrical system. The negative terminal of the 12 volt battery is connected by a large conductor to the chassis. This terminal is called "ground" even though it is insulated from earth by the rubber tires.

Electrical accessories in an automobile are connected to the positive 12 volt terminal of the battery through appropriate fuses with a single conductor connected to each accessory. Accessories are mounted securely to the chassis, since the chassis serves as the second (ground return) conductor to complete the circuit. The need for additional conductors is thereby eliminated.

In a similar manner, an electronic system has a large conducting area in a circuit which is the reference for all voltages in the system. This section, called the ground or the system ground bus, could be a metal chassis or metal enclosure for the system. It can also be a copper section on a printed circuit board or a separate metal bar mounted in a specific location in the system. Ground is designated graphically on a schematic drawing by one of the symbols shown in Figure 7.3 .

Ground Symbols
Figure 7.3

THE SINE WAVE
The movement of a swinging pendulum can be compared to the movement of alternating current (AC). Both move in the form of a unique wave shape called a *sine wave* .

When viewed from its center position, a moving pendulum moves in one direction for a certain distance and returns to its starting point. It then continues its movement in the opposite direction for the same distance and returns to the starting point. (See Figure 7.4)

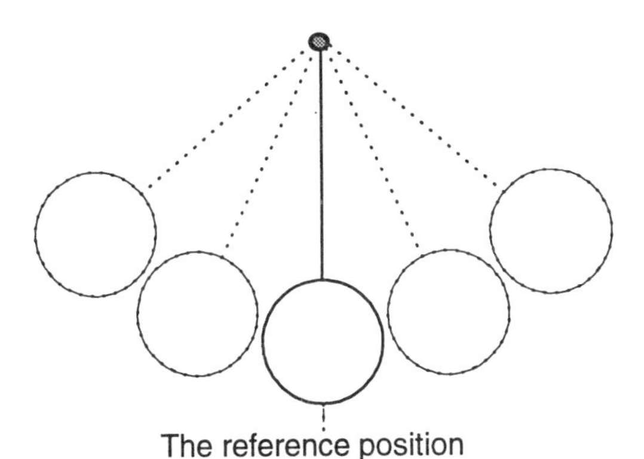

Action of a Moving Pendulum
Figure 7.4

The center position of the moving pendulum is designated as its reference. The action of the pendulum starts from the reference, swings in one direction, swings back to the center, swings in the opposite direction, and then back to the reference position to constitute one complete cycle. If there is no friction to slow down and, eventually, stop the pendulum, it would keep swinging back and forth continuously repeating each cycle indefinitely.

A graphic representation of a circuit is always shown with respect to a horizontal reference. The example of a pendulum at rest can be shown with its reference line shifted from its vertical to its horizontal plane.

To move the pedulum above and below the new reference, a mechanical force is exerted on it, first in an upward direction, then downward, and then back to the reference position. This continuing cyclical movement of the pendulum illustrates an alternating pattern called a sine wave.

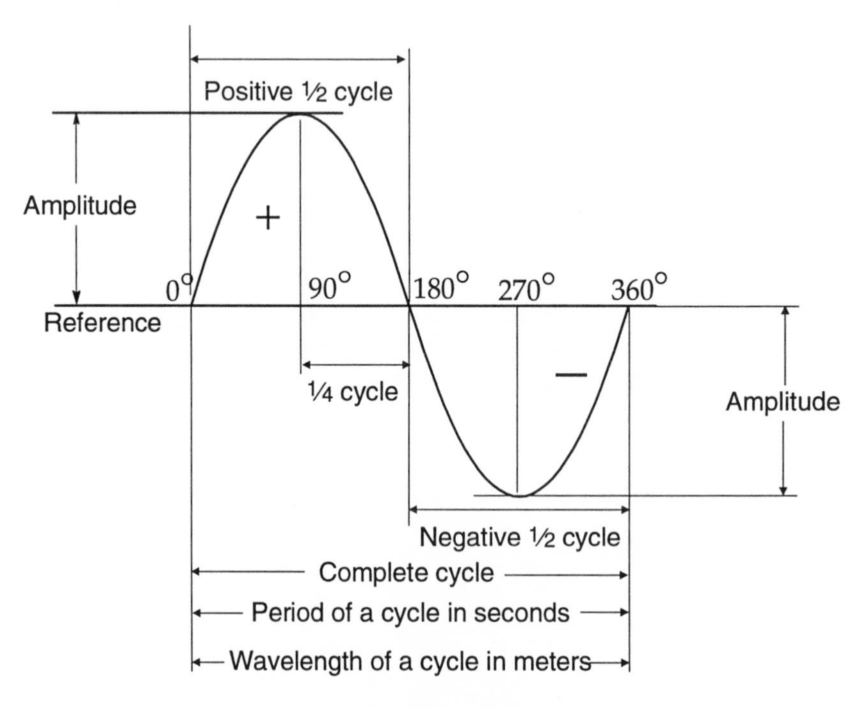

The Sine Wave
Figure 7.5

Each single-cycle of the sine wave has a positive and negative half with respect to its horizontal reference. Once a reference is established, the phrase "with respect to the reference" can be omitted since it is commonly assumed to exist.

A complete sine wave is continuous and smooth with no discontinuity or break in its appearance. The height of each half-cycle above and below the reference is called its *amplitude*. Both positive and negative half-cycles amplitudes are equal to each other. The sine wave is referred to as a symmetrical waveshape.

FREQUENCY, PERIOD, AND WAVELENGTH

As each cycle of the sine wave keeps repeating, the element of time must be taken into account. The number of cycles that occur in one second is called its *frequency*. The *hertz* (Hz) is the unit of measurement of the frequency and is named after **Heinrich Hertz**, a German physicist, who researched wave propagation during the latter part of the 19th Century.

The *period* of the waveshape is the length of time needed to complete one cycle and is designated by the letter **T**. The period is measured in fractions of seconds and is the reciprocal of the frequency, or 1 divided by the frequency.

$$\text{Period} = \mathbf{T} = \frac{1}{\mathbf{f}} \quad \text{in seconds, with the frequency, } \mathbf{f}\text{, in hertz}$$

If the period is known, this equation can be transposed to calculate the frequency of a sine wave. Frequency is the reciprocal of the period, or 1 divided by the period, and is expressed as:

$$\text{Frequency} = \mathbf{f} = \frac{1}{\mathbf{T}} \quad \text{in hertz, with the period, } \mathbf{T}\text{, in seconds}$$

As frequency increases, the length of time for one cycle decreases. For example, the period of a 1 kilohertz sine wave is 1 millisecond (1 ms). A sine wave with a frequency of 1 megahertz has a period of 1 microsecond (1 μs), etc.

Wavelength is the distance from the beginning to the end of one sine wave and is designated by the Greek letter, λ (lambda). Wavelength is measured in meters, or fractions of meters (centimeters, millimeters, micrometers, or nanometers).

As frequency increases, wavelength decreases. When the wavelength decreases to an extremely small value, as in light frequencies, it is sometimes measured in *Ångström units* (Å), an older and generally obsolete dimension. One Ångström unit, is equal to one-tenth of a billionth of a meter (one-tenth of a nanometer).

The Ångström unit, named after **Anders Ångström** of Sweden, a 19th Century physicist, has been replaced by the nanometer, the universally accepted dimension for wavelength.

Below the frequencies of light, a sine wave is usually specified by its frequency, **f**, and when indicated, can be designated by either its period, **T**, or its wavelength, λ, instead of its frequency.

The sine wave is an example of a waveshape consisting of a *single* frequency that is called its *fundamental* frequency. A complex waveshape is identified as having many frequencies that can be generated in a variety of ways.

If a taut guitar string is plucked, its vibrations cause molecules of air to vibrate, producing a fundamental frequency plus additional frequencies. These additional frequencies are multiples of the fundamental and are called *harmonics*. The 2nd, 4th, 6th, etc. are called *even harmonics*. The 3rd, 5th, 7th, etc. are called *odd harmonics*. The amplitude of each successive harmonic generally diminishes compared to the amplitude of the fundamental.

The sine wave, with no harmonics, will be used as the model for all references to AC in this text.

Another complex waveshape, called a *square wave*, is composed of the fundamental frequency plus an infinite number of odd harmonics. (See Figure 7.6) It is a rectangular-shaped periodic wave with a positive and negative half-cycle of equal duration, or width. The transition time between positive and negative halves is negligible compared with the duration of each half-cycle.

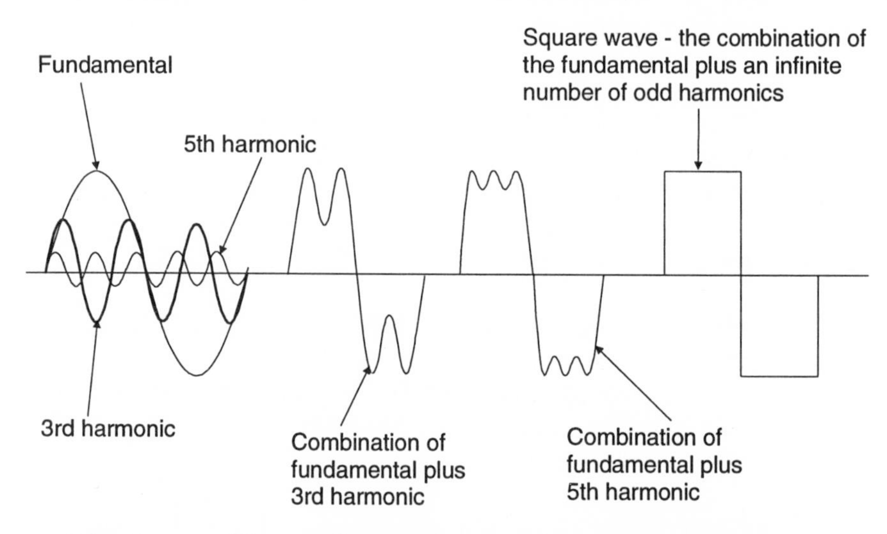

The Square Wave - The Fundamental Plus Odd Harmonics
Figure 7.6

The closer a square wave is to the ideal (zero rise and fall times), the more odd harmonics are present in the wave, approaching an infinite number of odd harmonics.

When current in a circuit is reversing, or alternating, every half cycle in a sine wave manner, it is called **alternating current** (AC). The voltage that produces this current is called an **AC voltage** and is specified by its amplitude, its frequency, and its current and power capabilities.

The waveshape of an AC supply voltage varies in a sine wave manner. As the AC voltage changes in this manner, from zero to a positive (plus) value, back to zero, and then to a negative (minus) value, the current produced in a circuit will change in the same way.

Small-signal AC voltages generated from phono cartridges, microphones, tape heads, radio antennas, etc., are also referred to as AC. These signals exhibit inherent sine wave characteristics.

The prime source of electrical power generated throughout the United States and the rest of the Western Hemisphere is nominally 120 volts AC at a frequency of 60 hertz and is referred to as the *line voltage* or the *power line voltage*. In Europe, the nominal line voltage is 240 volts AC at a frequency of 50 hertz.

The AC line voltage normally used in airborne and space applications is generated at a frequency of 400 hertz. This higher frequency permits the use of smaller and lighter components in systems used in these applications, where space and weight are critical parameters and need to be minimized.

THE FREQUENCY SPECTRUM

A *frequency spectrum* indicates the specific range of frequencies generated, or radiated, and specified for a variety of applications.

There are two general types of radiation:

- Sonic wave
- Electromagnetic wave

SONIC (ACOUSTIC) RADIATION

Sonic waves are mechanical vibrations that travel 1116 feet (340 meters) per second in air at sea level. (See Figure 7.7)

The categories included within the sonic group are:

Subsonic - frequency range between 10ths of a hertz to 20 Hz - used in seismographs and sonar detection systems

Audio (the frequency range of human hearing) - nominally between 20 Hz and 18 kHz

Ultrasonic (supersonic) range - frequency range is beyond the audio range, typically from 180 kHz to 1MHz and above. These frequencies are used in applications, including: ultrasonic cleaning, ultrasonic bonding (for attaching gold leads between a silicon chip and its package terminals), and sonogram scanning in medical diagnostic equipment. Ultrasonic scanners are used in the aviation and aerospace industry to test metallic materials for fatigue or damage.

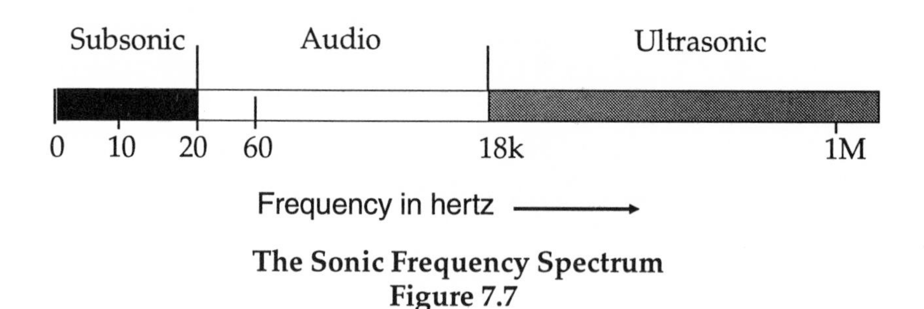

The Sonic Frequency Spectrum
Figure 7.7

ELECTROMAGNETIC RADIATION (EMR)

Electromagnetic energy is radiated within electrical and electronic circuits that are producing oscillations or sine waves that travel at the speed of light - 186,000 miles (3×10^8 meters) per second. This type of radiation includes:

- **Extremely Low Frequency** (ELF) - from 3 Hz to 3 kHz - is used in specifying computer monitor radiation parameters and radiation of electromagnetic waves caused by the 60 Hz power line

- **Radio Frequency** (RF) - These frequencies range between 3kHz and 300 GHz and are commonly used as the *carrier waves* in radio communications systems to transmit (carry) information such as audio, video, and digital information for long distances at the speed of light. The RF spectrum is divided into eight frequency bands. Each succeeding band is ten times as high in frequency as the one just below it in the spectrum.

The RF bands include:

Very Low Frequency (VLF) - 3 kHz to 30 kHz - restricted bandwidths within this portion of the spectrum prevent its use for communications, although proposals have been made for worldwide communication use in submarines.

Low Frequency (LF) - 30 kHz to 300 kHz - used for radio telegraphy with few interruptions from diurnal, seasonal, or solar interference

Medium Frequency (MF) - 300 kHz to 3 MHz - used for radio broadcasting, maritime and aeronautical radio communications, radio navigation, and amateur radio communications

High Frequency (HF) - 3 MHz to 30 MHz - used for mobile services, commercial radio broadcasting, maritime, amateur and CB radio communications, and telemetering

Very High Frequency (VHF) and **Ultra High Frequency** (UHF) - 30 MHz to 3 GHz - used for relatively short-distance communications, including TV, radar, space research, radio astronomy, and amateur radio communications

Super High Frequency (SHF) and **Extremely High Frequency** (EHF) - 3 GHz to 300 GHz - microwave region - used for microwave ovens, satellite communications, meteorological aids, high-definition radar, and radio frequency spectroscopy

Light Frequencies begins above the EHF range. Measurements are specified in wavelength (nanometers). Nonvisible infrared lies in the range between 4,000 to 700 nanometers. The spectrum ascends into the visible light range, going from red (700 nanometers) to violet (400 nanometers), and then to nonvisible ultraviolet (from 400 to 12 nanometers).

X-Ray Region - 10 to 1 nanometer - used extensively in medical, biological, and industrial applications. The use of X-rays has contributed to the understanding of the structure of matter

Gamma Ray Region - 0.14 to 0.001 nanometers - used in the study of nuclear energy

Cosmic Ray Region - 0.1 to about 0.01 picometers - completing the electromagnetic radiation frequency spectrum

DIRECT CURRENT (DC)

AC voltage is the source of electrical power for lights, heaters, motors, and similar electrical machinery. These circuits and systems that use AC as their primary source voltage are categorized as "electrical" circuits or systems. Other circuits and systems classified as "electronic" require steady-state DC voltage as their source of power. This voltage source is used for radio, computers, calculators, amplifiers, TV, and similar electronic systems.

DC differs from AC in that DC does **not** have alternating polarities with respect to a reference. DC maintains a constant polarity, either plus or minus, with respect to a fixed reference.

There are three forms of DC voltage:
- Pulsating DC
- Steady-state DC
- Pulsed DC (digital pulses)

A battery is one source of steady-state DC voltage and works very well for portable systems, such as hand-held calculators, digital watches, and tape recorders. For systems consuming substantial amounts of power, batteries will have a very short life span and their use would be most impractical for large, non-portable, high power-consuming systems.

The AC power line is used as the initial, but not the final source of power, since it has the capability of providing sufficient power for non-portable electronic systems that cannot be operated practically with batteries. For these applications, the AC voltage from the power line must be converted to the required steady-state DC voltage.

PULSATING DC VOLTAGE - HALF WAVE

The first step in the conversion of AC voltage into steady-state DC is to change the AC into *half-wave* or *full-wave pulsating* DC.

Half-wave pulsating DC voltage is produced by removing half the AC sine wave allowing current to flow only in one direction, resulting in either a negative or positive voltage at the output. This is called *half-wave rectification* and produces a waveshape that has a single voltage polarity. (See Figures 7.8 and 7.9)

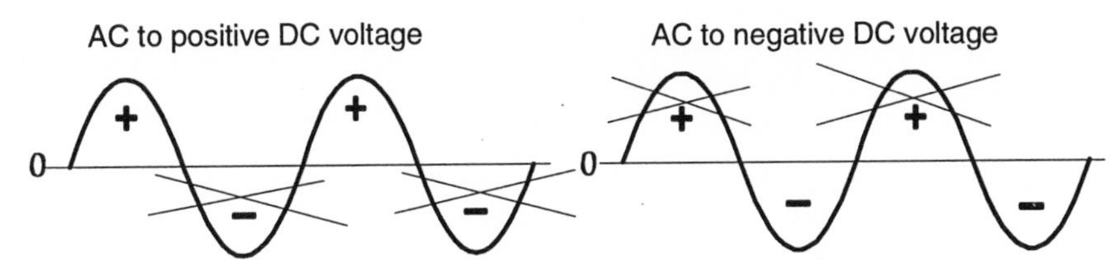

AC Voltage to be Changed to Pulsating DC
Figure 7.8

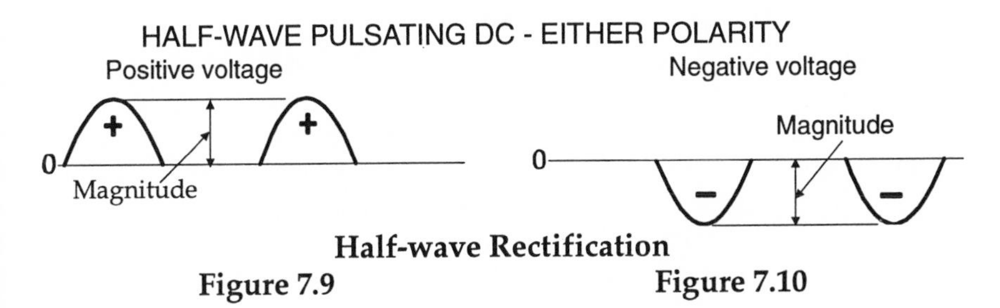

Half-wave Rectification
Figure 7.9 **Figure 7.10**

Depending on the way a rectifier circuit is configured, the resultant pulsating DC voltage has either a positive (Figure 7.9) or a negative (Figure 7.10) voltage polarity with respect to a reference. The half-wave pulsating DC voltage that is produced is satisfactory for some DC voltage supply applications, but in some cases, half wave rectification does not provide the most useful means of rectification, since only half of the AC sine wave is used.

PULSATING D.C. VOLTAGE - FULL WAVE

A more efficient approach to rectification is to create full-wave pulsating DC voltage through *full-wave rectification*. (See Figures 7.11 and 7.12) This is accomplished by "flipping" half of the sine wave, removed by half-wave rectification, to the other side of the reference.

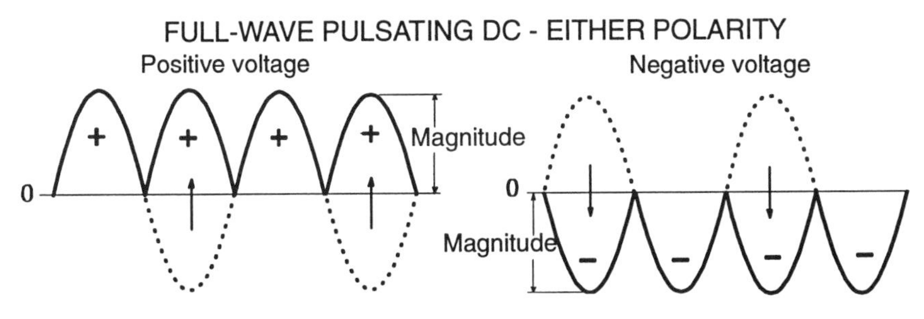

Full-wave Rectification
Figure 7.11 **Figure 7.12**

- For positive voltages, the negative half of the sine wave is flipped to the positive side. (See Figure 7.11)

- For negative voltages, the positive half of the sine wave is flipped to the negative side. (See Figure 7.12)

The process of rectification, both half-wave and full-wave, is discussed in detail in Volume Two - Part I - Discrete Semiconductors.

STEADY-STATE D.C. VOLTAGE

After rectification, pulsating DC voltage is changed into steady-state DC voltage with the use of one or more *filter capacitors* that serve to remove the pulsations and maintain a ripple-free level of DC voltage. Filtering is covered in detail in Chapter Ten - Capacitors.

Steady-state DC voltage, shown in Figure 7.13, is specified by its magnitude and polarity and could be either positive or negative.

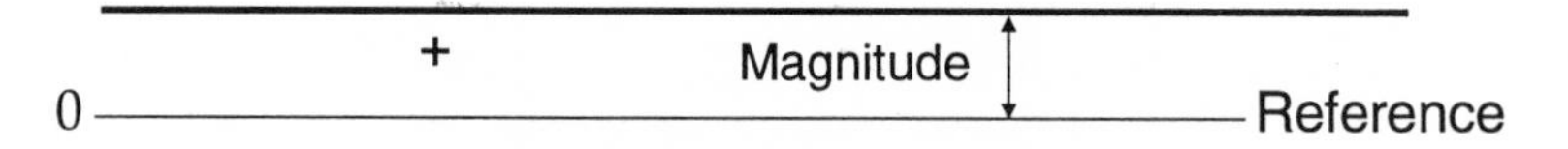

Steady-state DC Voltage
Figure 7.13

Generally, steady-state DC voltage is held at a constant level with respect to a reference. The process of maintaining a constant level of voltage (constant magnitude) is called *regulation* regardless of any changes in line voltage or in the resistance value of a load.

REGULATION

The nominal 120 volt AC line is generally held to within ± 1 volt, however, the line voltage in some areas can be a source of *unregulated*, or *varying* voltage. The line voltage can vary as much as ± 15% or more, which may be undesireable for some electronic circuits.

If this condition exists, a relatively simple method of voltage regulation is accomplished electronically within the power supply section of the electronic circuit after the steps of rectification and filtering are completed.

Electronic voltage regulation using a zener diode is described in detail in Volume Two - Part I - Discrete Semiconductors.

PULSED DC VOLTAGE (DIGITAL PULSES)

Pulsed DC voltage, or *digital pulses* are used in electronic circuits that make up the structure of systems used in calculators and computers. They are fixed polarity pulses (either positive or negative) that start at the reference, or base line (zero), increase instantly to a specified height, or magnitude, stay at that value of voltage for a specified time, and then instantly return to zero.

An ideal digital waveshape exhibits zero transition time from zero to the maximum height of the voltage pulse and zero transition time on returning to zero (the reference). This voltage is referred to as a *digital,* or *step-function pulse,* since it describes a sudden, or instantaneous change in voltage. The change is either *step-up,* or *step-down,* from a reference to a specified magnitude, and, eventually, its return to the reference (0 volts), indicating a digital pulse that is either positive or negative with respect to the reference.

There are two types of digital pulses:
- Clock pulses
- Random pulses

Clock pulses provide the timing for computers and are specified by polarity, magnitude, *clock frequency,* and *duty cycle.*

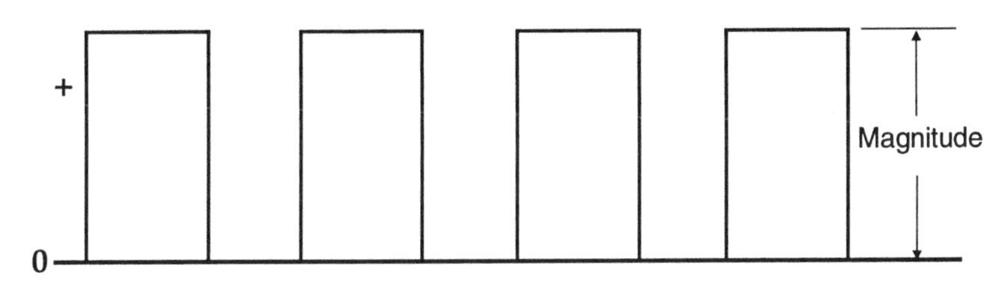

Clock Pulses
Figure 7.14

Clock frequency, also called *repetition rate,* specifies the number of digital pulses generated per second. Duty cycle is defined as the length of time that a pulse is ON, compared to the total time that the pulse is ON and OFF, expressed as a percentage of that total time.

For example, if a pulse is in its ON state for 1 microsecond and in its OFF state for 1 microsecond, the combined time of the ON and OFF pulses is 2 microseconds. In this case, the ON time is ½ the total time, and the clock pulse is specified, therefore, as having a 50% duty cycle.

Duty cycles can range from just above 0% to just below 100%. A voltage having a **duty cycle of 100%** is **steady-state DC**.

Every computer has an electronic clock as part of its circuitry which generates digital pulses similar to those shown in Figure 7.14.

Random pulses are digital pulses consisting of step-function voltages that have a magnitude and polarity but do not have a frequency (repetition rate) or duty cycle. (See Figure 7.15) They are specified in a schematic drawing, or in other descriptive literature, by their *pulse width* (PW) - the length of time each pulse is ON, their magnitude, and their polarity.

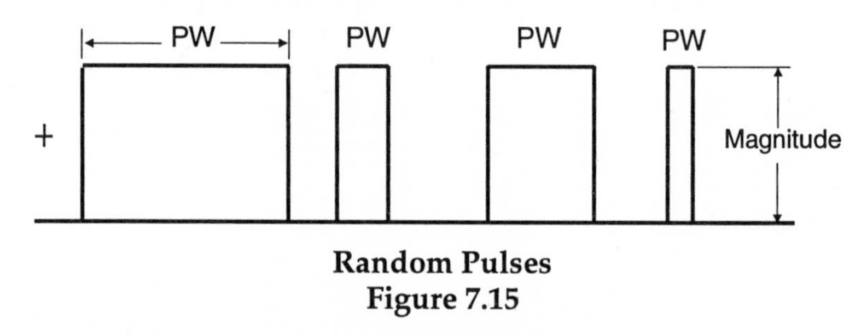

Random Pulses
Figure 7.15

These pulses are used for control functions in circuits, to turn equipment ON and OFF, or to suddenly change a circuit condition from one state to another. They can be patterned for use as the data or the *digital words* in a computer system.

Digital pulses and their use in digital circuitry are discussed in detail in Volume Three - Integrated Circuits and Computer Concepts.

CHAPTER
EIGHT

GENERAL INFORMATION

THE EXPONENTIAL SYSTEM

THE METRIC SYSTEM

PASSIVE AND ACTIVE COMPONENTS

DISCRETE COMPONENTS AND ICs

POWER AND SMALL-SIGNAL CATEGORIES

TEMPERATURE CONSIDERATIONS
- TEMPERATURE SCALES - FAHRENHEIT AND CELSIUS
- OPERATING AND STORAGE TEMPERATURE RANGES
- TEMPERATURE COEFFICIENT

TRANSDUCERS

GENERAL INFORMATION

THE EXPONENTIAL SYSTEM

When large and small numbers need to be written in an abbreviated form, the *exponential system* is used. This system has a *base number* and an *exponent* to represent a specific numerical value. The exponent (*superscript*) is the number placed to the right and slightly above a base number. Both the exponent and the base number can have any value. (See Figures 8.1 and 8.2)

Example	Base Number	Exponent	Numeric Value	Comments
10^0	10	0	1	Expressed as: 10 to the zero
2^0	2	0	1	Expressed as: 2 to the zero
10^1	10	1	10	Expressed as: 10 to the 1
2^1	2	1	2	Expressed as: 2 to the 1
5^2	5	2	25	Calculated as: 5 x 5 or 5 multiplied by itself Expressed as: 5 squared or 5 to the 2
2^2	2	2	4	Expressed as 2 squared
10^3	10	3	1000	Calculated as: 10 x 10 x 10 Expressed as 10 cubed or 10 to the 3
2^3	2	3	8	Calculated as: 2 x 2 x 2 Expressed as: 2 cubed or 2 to the 3

Positive Exponents
Figure 8.1

Example	Base Number	Exponent	Numeric Value	Comments
10^{-1}	10	-1	0.1 or 1/10	Calculated as: 1 ÷ 10 Expressed as : 10 to the -1
2^{-1}	2	-1	0.5 or ½	Calculated as: 1 ÷ 2 Expressed as : 2 to the -1
10^{-2}	10	-2	.01 or 1/100	Calculated as: 1 ÷ 10^2 Expressed as: 10 to the -2
2^{-2}	2	-2	0.25 or ¼	Calculated as : 1 ÷ 2^2 Expressed as: 2 to the -2
10^{-3}	10	-3	.001 or 1/1000	Calculated as: 1 ÷ 10^3 Expressed as: 10 to the -3
2^{-3}	2	-3	.125 or ⅛	Calculated as : 1 ÷ 2^3 Expressed as: 2 to the -3

Negative Exponents
Figure 8.2

SIMPLIFIED CALCULATION FOR BASE 10

- For positive exponents, the value of the exponent indicates the number of zeros to be placed to the **right** of number 1.

- For negative exponents, the value of the exponent indicates the number of places the decimal point is to be moved to the **left**.

THE METRIC SYSTEM
MULTIPLES AND DIVISIONS OF 1000

The **metric system** is an international decimal system of weights and measures. In the electrical and electronic fields, standard metric system prefixes indicate multiples and divisions of 1000 for the measurements of: volts, ohms, amperes, watts, hertz, seconds, meters, and values of all electrical and electronic components. Metric prefixes are used with numbers greater than (>) 999 and less than (<) 1.

PREFIX	ABBREVIATION	EXPONENTIAL VALUE	MULTIPLY BY:
tera (trillion)	T*	10^{12}	1,000,000,000,000
giga (billion)	G*	10^{9}	1,000,000,000
mega (million)	M*	10^{6}	1,000,000
kilo (thousand)	k	10^{3}	1,000
			DIVIDE BY:
milli (thousandth)	m	10^{-3}	1,000
micro (millionth)	µ**	10^{-6}	1,000,000
nano (billionth)	n	10^{-9}	1,000,000,000
pico (trillionth)	p	10^{-12}	1,000,000,000,000
femto (quadrillionth)	f	10^{-15}	1,000,000,000,000,000

* The first letter of the measurement's abbreviation is capitalized for quantities of 1 million (mega) and higher.
** The abbreviation for micro is the Greek letter mu (µ).

Metric Notations
Figure 8.3

The metric system is used world-wide by scientists of all disciplines and for general purposes in nearly all countries of the world. At this writing, the United States is the only industrialized nation in the world that has not officially adopted this standard system of weights and measures, but is making progress toward that end. Nevertheless, metrics are used extensively in electrical and electronic component data sheets, specification control drawings, and technical manuals in the electrical and electronics industry in the United States.

PASSIVE AND ACTIVE COMPONENTS

Components are electrical or electronic devices that are assembled and interconnected to create circuits and systems. When connected to a supply voltage, components perform electrical or electronic functions and are generally classified as either **passive** or **active.**

> A **passive component** is a device that can sense, monitor, transfer, attenuate, vary, or control a voltage but does **not** differentiate between a positive or negative polarity and does **not** provide amplification to the voltage it is sensing. Amplification causes a signal or voltage applied to the input of a component to increase in amplitude or power at the output. A passive component performs additional functions, such as: voltage storage, switching, and timing.

Included among passive components are fixed resistors, variable resistors (potentiometers, rheostats and trimmers), capacitors, mechanical switches, electromechanical relays, inductors, transformers, crystals, and connectors.

In comparison, an **active component** does differentiate between a positive or negative voltage polarity by responding differently to each polarity. In many cases, it will also amplify the voltage it is sensing, monitoring, or controlling. Before 1954, *vacuum tubes* were used as the active components in circuits and, except for some highly specialized types, have been made obsolete by **semiconductors**.

Included in the semiconductor category are diodes, thyristors (silicon controlled rectifiers and TRIACs), bipolar transistors, and field effect transistors (FETs). These components are discussed in Volume Two - Part I - Discrete Semiconductors.

DISCRETE COMPONENTS AND ICs

Electronic components are further categorized according to the manner in which they are packaged or made available for use in circuitry. The finished device can be either in the form of a *discrete component* or packaged as part of a *hybrid integrated circuit* (IC).

- A discrete component is a single component in a finished package or in chip form. This category includes both discrete passive components and discrete semiconductors.

- An integrated circuit can be either in hybrid or monolithic form.

- A hybrid IC can contain passive components (resistors and capacitors), discrete semiconductors (diodes and transistors), monolithic ICs, or any appropriate combination thereof. These components are attached to a pre-wired, single-layer, or multi-layered *substrate* to make up the desired circuit, group of circuits, or an *array*.

  A substrate is a material that provides the supporting, interconnecting, and insulating base for the hybrid circuit components. Substrate material includes: ceramic, alumina ceramic, and silicon. The final hybrid assembly is then enclosed in a package having external terminals that are connected to an external system.

  An array consists of a group of similar components (e.g., a resistor array, a diode array, etc.) assembled in a package.

- A *monolithic IC* is a semiconductor structure processed on a single, silicon (Si) or gallium arsenide (GaAs) substrate into which various chemicals (dopants) have been diffused. When in vapor form, these dopants act to change the characteristics of the silicon or gallium arsenide substrate to produce the functions of resistors, capacitors, diodes, transistors, conductors, and non-conductors on the substrate (referred to as an *active substrate*).

  The basic monolithic IC manufacturing process is a metallurgic, photolithographic, and chemical technique that can create one circuit, many circuits, an array, a small or large subsystem, or a small or large total system on a single chip. Processing is done on a wafer containg many identical chips that are initially tested on the wafer and then separated. The individual chip is then assembled inside a package with external terminals that connect the circuit or system on the chip to external circuitry.

For both hybrid and monolithic IC technologies, the final, single package is assigned a part number that designates a total circuit, system, or entity, without identifying or specifying the characteristics of its individual components or component functions.

The subject of integrated circuits, both hybrid and monolithic, is discussed in detail in Volume Three - Part I - Integrated Circuits.

POWER AND SMALL-SIGNAL CATEGORIES

If either AC or DC sources have a power capability of 3 watts or more, they are classified as *power* sources. If the sources provide less than 3 watts, they are called *small-signal*. These categories also apply to discrete components and integrated circuits.

TEMPERATURE CONSIDERATIONS

TEMPERATURE SCALES - FAHRENHEIT VS. CELSIUS

In 1714, **Gabriel Fahrenheit**, a German-Dutch physicist, designed the first practical mercury thermometer using a scale having increments of 180° between the freezing point (32° F) and boiling point (212° F) of water at sea level.

In 1743, **Anders Celsius**, a Swedish astronomer, in attempting to simplfy the divisions of the Fahrenheit scale, developed a scale that divided the range between the freezing and boiling points of water at sea level into an even 100°. Because of the ease in working with decimal increments, this scale is the preferred method of measuring temperature in the modern scientific world.

The scale first developed by Celsius used 0° as the boiling point of water and 100° as the freezing point. In 1744, the scale was changed to the present method (freezing water = 0° and boiling water = 100°) and was named the *centigrade* scale. In 1948, to honor its inventor, it was renamed the *Celsius scale*. Electrical and electronic component data sheets and equipment manuals describing electrical and electronic systems generally use the Celsius scale for temperature specifications. See Figure 8.4 for conversion between the two scales.

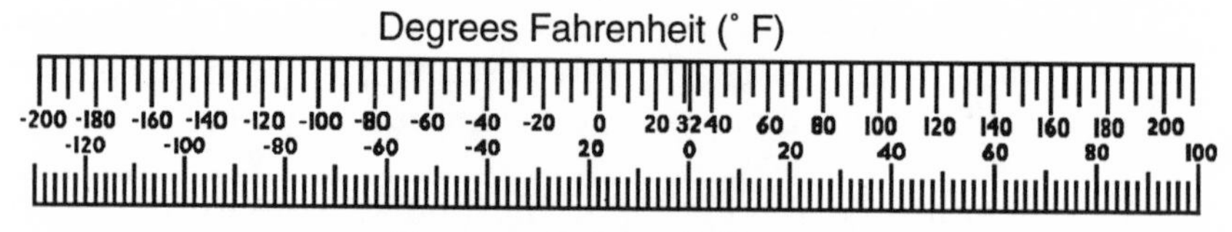

Figure 8.4

$$\text{Degrees Celsius (° C)} = (\text{Degrees Fahrenheit} - 32) \times \frac{5}{9}$$

$$\text{Degrees Fahrenheit (° F)} = (\text{Degrees Celsius} \times \frac{9}{5}) + 32$$

OPERATING AND STORAGE TEMPERATURE RANGES

All components are specifed to operate and be stored within a defined temperature range, however, the operating and storage ranges need not be the same. Depending upon whether they are intended for military and space use, or for commercial, industrial, and consumer applications, components must conform to their specified temperatures ranges:

- **Military and space applications**: From -55° C to +125° C

- **Commercial, industrial, and consumer applications**: This range is not specifically defined. Depending on the end use, the temperature could have any specified limits. A common range is from 0° C to +70° C, however, these limits are not fixed, but depend on specific requirements. Some typical ranges are: +10° C to +85° C, +10° C to +75° C, 0° C to +80° C, etc.

TEMPERATURE COEFFICIENT - TC

The **temperature coefficient** (TC) is defined as the change in the nominal value of a component for every °C change in temperature. The characteristics of components are generally specified at a *reference*, or *nominal*, temperature of +25° C (+76° F). This temperature refers to the normal room temperature of an indoor laboratory and is used as the reference in specifying the nominal characteristics of components.

The TC must be considered in the specification of components and the design of equipment. Since the temperature in electronic systems normally varies from the reference, or nominal value, the characteristics of a component will change as the temperature in and around it changes. The temperature coefficient could have either a positive temperature coefficient (PTC) or a negative temperature coefficient (NTC).

- PTC - As temperature increases, the component's value increases according to its temperature coefficient, exhibiting a *direct relationship* between the changing temperature and the change in the component's value.

- NTC - As temperature increases, the component's value decreases according to its temperature coefficient, exhibiting an *inverse relationship* between the changing temperature and the change in the component's value.

TRANSDUCERS

A **transducer** is a device, component, circuit, or system that converts one form of energy into another.

A thermostat may react to changes in temperature by providing the electrical analog, or voltage equivalent, of the changing temperature. This voltage can be used to control the operation of a heating or cooling unit.

Another type of transducer, a record-player cartridge, accepts mechanical vibrations of a stylus moving over a phonograph disk and converts the mechanical energy produced by that movement into equivalent electrical voltages. These voltages are then amplified by an electronic circuit connected to a loudspeaker. As another type of transducer, the loudspeaker converts the electrical energy from the amplifier into mechanical vibrations to fill a room with sound.

Some types of transducers convert electrical energy into other forms of energy, while others convert other forms of energy into electricity. Whichever transducer is used, it can be either passive or active in nature and could be a single element or a complex circuit with many components forming a single transducer system. Transducers are manufactured in a great variety of packages. They have many different electrical and mechanical characteristics and unique features, performing a vital function in many electronic systems.

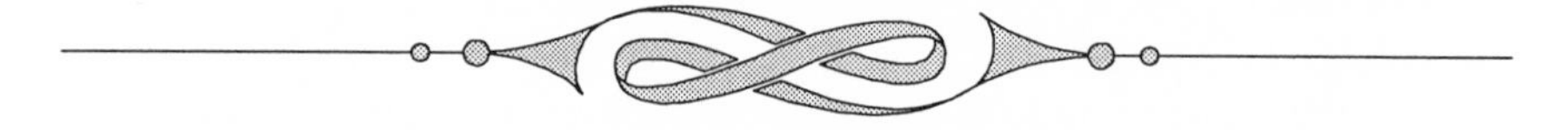

The basic concepts and rules covered in Part One are the foundation for all the succeeding information in this series of books. These fundamentals provide the alphabet, vocabulary, and structure of electrical and electronic circuits. The world of circuits, and systems, however, consists of tangible components that are put together according to these concepts and rules.

In the succeeding sections of the series, electronic components are discussed to show how they are made, how they are specified on a data sheet, how they function, their specific applications in circuitry, and how they interrelate with other components to create electrical and electronic circuits and systems.

APPENDIX

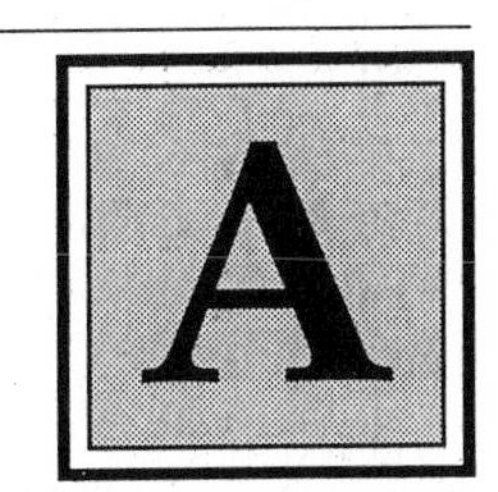

REINFORCEMENT EXERCISE

BASIC CONCEPTS

Answer TRUE or FALSE

1. Ohm's Law defines the relationship between the voltage, E, in volts, the circuit resistance, R, in ohms (Ω), and the resultant current flow, I, in amperes. To calculate one unknown value, the other two values must be known.

2. With the resistance in a circuit remaining constant, current will increase if the supply voltage increases and will decrease if the supply voltage decreases.

3. As the resistance in a circuit decreases, and the supply voltage, E, remains constant, the total current flow in the circuit will decrease in accordance with Ohm's Law: $I = E \times R$.

4. A device, component, appliance, or machine in an electrical or electronic circuit which uses the available energy of the supply voltage (voltage source) is called a load. At any one time, this load has a specific resistance that is measured in ohms.

5. As more resistance or electrical devices are added in series in a circuit, the total resistance of the circuit is increased. The current in the circuit will decrease if the supply voltage is held constant.

6. In a series circuit, the sum of the individual voltages external to the supply voltage is equal to the supply voltage.

7. As more resistance or electrical devices are added in parallel, the total or equivalent resistance of the circuit will increase. The total current will decrease if the supply voltage is held constant.

8. In a parallel circuit, all voltages across the supply and across all the loads are equal to each other in both amplitude (magnitude) and polarity.

9. A conductor is defined as a material having essentially zero resistance and is shown as a straight line in a schematic drawing. A non-conductor is defined as a material having essentially infinite resistance and has no graphic symbol.

10. Copper is the best material for use as a conductor because it has the lowest resistance (highest conductivity) of all conductive materials.

11. The function of a fuse or circuit breaker is to provide a means of protecting the components of a circuit, the conductors in the circuit, and the voltage source.

12. The choice of using either a fuse or circuit breaker as a circuit protective component depends on the initial cost, the amount of future maintenance, and convenience in its use. A fuse costs less but requires replacement when blown. A circuit breaker merely requires a re-set operation to restore it to its protective mode.

13. In a single-load circuit, power consumed by the load is equal to the power generated by the source voltage which is equal to the source voltage times the load current ($P = E \times I$).

14. Doubling the voltage across a fixed load will result in doubling the power consumed by that load.

15. Assuming no losses, the total power consumed in a circuit is equal to the sum of the power consumed by each load, regardless of whether the circuit is a series or parallel configuration.

16. The heat generated in a component (I^2R in watts) has no effect on the component or on any other components in its vicinity. Heat within a circuit is not the circuit designer's concern.

17. As the frequency of a sine wave increases, both the period of the sine wave and its wave length increases.

18. AC voltage is preferred as the final voltage source for the electronic circuits in the home since it is much less expensive than DC voltage.

19 DC voltage can be either half-wave or full-wave pulsating DC, steady-state DC, or pulsed DC (digital pulses).

20. Steady-state DC must be used as the final source of voltage in electronic equipment, such as: amplifiers, computers, instrumentation systems, radio, and television.

Calculate the voltage across each load, the current through each load, the power consumed in each load, and the total power consumed in the following circuits:

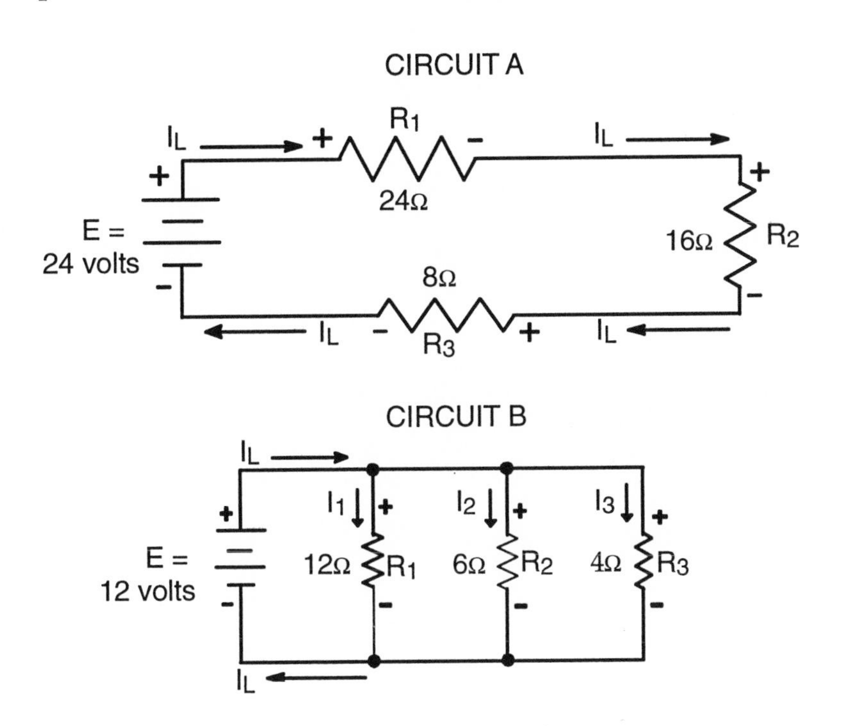

Calculate the frequency of the sine wave shown below:

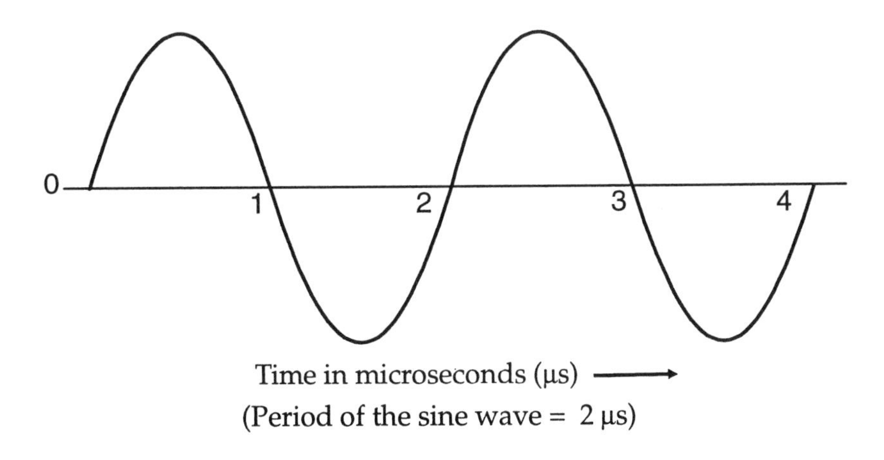

Answers to this reinforcement exercise are on pages 115-121

APPENDIX

GLOSSARY OF POPULAR ELECTRONIC TERMS

ALTERNATING CURRENT - AC VOLTAGE - Either voltage or current that varies smoothly from zero to a maximum value in one direction, or polarity and then returns to zero. It then reverses its direction (polarity) and rises to a maximum value in the opposite direction, and then returns to zero to complete its cycle. This cycle is repeated continuously. The number of cycles per second is its frequency, measured in hertz (Hz). See SINE WAVE.

AMPERE - The unit of measurement of electrical current flow, named after André Ampère, a 19th Century French physicist. One ampere is the value of current that will be maintained in a circuit with an electromotive force of one volt and a resistance of one ohm. One ampere = 6.25×10^{18} electrons/second. See CURRENT.

AMPLIFICATION - The process of increasing the voltage, current, or power of an electrical signal.

AMPLIFIER - An electronic circuit that draws power from a supply voltage, or voltage source, to produce, at its output, an increased reproduction of the signal existing at its input. The amplifying component could be a transistor, vacuum tube, or an appropriate magnetic device.

ANALOG VOLTAGE - A gradually changing voltage. The term is interchangeable with LINEAR VOLTAGE. For example, the voltage sensed by an automobile's speedometer is the analog of the speed of the automobile.

ÅNGSTRÖM UNIT - A unit of length that measures wavelength and is equal to 0.1 of a billionth of a meter (1×10^{-10} meters). It is named after Anders Ångström, a 19th Century Swedish physicist.

ARMATURE - The moving part of a magnetic device consisting of one or more coils that are electrically connected to create the rotatable section of a generator. See ARMATURE in Glossary of Switches, etc.

BATTERY - An electrical device consisting of one or more cells which converts chemical or solar energy into electrical energy. A battery provides a source of steady-state DC voltage.

BLOW TIME - The maximum time required for a fuse to open after being subjected to an excess of the device's rated current. Fuses are classified by blow time as slow, normal, or fast.

CELL - A single unit of a battery which generates a DC voltage or current by converting chemical or solar energy into electrical energy.

CIRCUIT - A single or group of interconnected components powered by a source of voltage and is configured according to specified rules. A circuit performs a specific or a general task that is usually predetermined.

CIRCUIT BREAKER - An automatic magnetic or bimetallic device that will open a current-carrying circuit under a condition of excessive current without being damaged. Unlike a fuse that melts when its rating is exceeded, a circuit breaker can be reset automatically, or manually, when the circuit problem is eliminated.

COIL - A length of insulated wire wound around a core. It may also be a free-standing device with air as the core.

COMPONENT - An individual part or element of an electrical or electronic circuit which performs a designated function within that circuit. It may consist of a single part or a combination of parts or assemblies.

CONDUCTOR - A metal material that carries electrical current and has essentially zero resistance.

CURRENT - The movement of electrons per second through a conductor or a component . It is measured in amperes and is designated by the letter, I. There are 6.25×10^{18} electrons per second in one ampere. (10^{18} = a billion billion)

DECAY TIME - The time it takes for a voltage to be reduced to a given percent of the peak voltage. See ELECTROSTATIC DISCHARGE.

DELAY TIME - The time it takes for a circuit breaker to open after its rated current is exceeded. See BLOW TIME.

DIGITAL VOLTAGE - A discontinuous or step-function electrical pulse characterized by an instantaneous change from zero to some finite level, either in a positive or negative direction with respect to a reference.

DIRECT CURRENT - DC - An electrical current or voltage with a constant direction (polarity) with respect to a fixed reference. DC can be either positive or negative.

ELECTRICAL GENERATOR - An assembly consisting of a magnet, mounted on a frame, and a wire coil (armature) that can be rotated within the magnetic field. The function of the generator is to convert mechanical energy into electrical energy. See TURBINE.

ELECTROMOTIVE FORCE (EMF) - The electrical force that exists across the terminals of an electrical generator, or battery. When connected to a load in a closed circuit, this force produces a voltage across the load and causes current to flow in that circuit. EMF is measured in volts and designated with the letter E (supply voltage) or V (load voltage).

ELECTRON - Considered to be the smallest unit of electrical charge.

ELECTROSTATIC CHARGE - The accumulation of electrons on the surface of a nonconducting material when it is rubbed by a different nonconductive material. See TRIBOELECTRIC EFFECT.

ELECTROSTATIC DISCHARGE (ESD) - A transfer of an electrostatic charge between a material having an excess of electrons to a material with a deficiency of electrons.

EXPONENTIAL SYSTEM - A convenient means of writing large and small numbers using a base number and a superscript, e.g., 10^2.

FREQUENCY - The number of cycles per second of an AC wave measured in hertz (Hz) and designated with the letter f.

FUNDAMENTAL FREQUENCY - The lowest frequency of a complex AC waveshape, represented by a single sine wave.

FUSE - A short strip of metal having extremely low resistance and functioning as a protective device in a circuit. A fuse will melt when its rated current is exceeded, thereby opening the circuit .

GROUND - The part of a circuit or system which is used as the reference for the voltages existing in that circuit or system. The ground consists of a conductive material such as copper, steel, or aluminum. See REFERENCE.

HARMONICS - Multiples of a single sine wave (the fundamental frequency). The even harmonics are the 2nd, 4th, 6th, etc., and the odd harmonics are the 3rd, 5th, 7th, etc. All harmonics are multiples of their fundamental frequency.

HEAT SINK - A metal base or plate onto which one or more components are mounted to absorb, carry away, or radiate the heat generated by the component(s). Overheating may result in the malfunction or destruction of the part(s) generating the heat or might cause damage to other parts of the circuit.

HERTZ (Hz) - The unit of measurement for the frequency of a sine wave or square wave, named after Heinrich Hertz, a 19th Century German physicist. The term hertz designates the number of cycles per second exhibited by these waves.

HORSEPOWER (HP) - A unit of measurement of mechanical power. It indicates the ability of a device or mechanism to do a specific amount of work over a period of time. It is equal to 550 foot-pounds per second in mechanical power or 746 watts in electrical power.

LIGHTNING ARRESTOR - A protective device that provides a very low resistance path to any voltage above its rated value. See METAL OXIDE ELEMENT.

LINE VOLTAGE - The AC voltage supply that provides the prime source of electrical power for office, laboratory, factory, and home electrical and electronic equipment. Throughout North, Central, and South America, the line voltage is nominally specified as 120 volts AC, at 60 hertz. In Europe, the line voltage is nominally specified as 240 volts AC, at 50 hertz. Line voltage can be either privately or publicly generated.

LOAD - A device, component, appliance, system, or machine to which an electrical force (voltage) is applied. Resistance is inherent in the structure of a load and is an integral part of an electrical or electronic circuit.

METAL OXIDE ELEMENT - A resistive device used to protect against excessive voltage surges in a circuit, generally against lightning. It is sometimes called a metal oxide varistor (MOV). Below its rated voltage, its extremely high resistance has no effect on a circuit. Above its rated voltage, it sharply changes to an extremely low value resistor. See LIGHTNING ARRESTOR.

NONCONDUCTOR (INSULATOR) - A material that has essentially infinite resistance (generally greater than 10^{10} ohms). It protects the circuit by isolating components and conductors from each other to prevent the possibility of a "short circuit".

OHM - The unit of measurement of resistance symbolized with the Greek letter omega (Ω). It is named after Georg Ohm, a 19th Century German physicist. One ohm is the value of resistance through which an electromotive force of one volt will maintain a current of one ampere. See RESISTANCE.

OHM'S LAW - The relationship that exists between the electrical parameters of voltage (electrical pressure), resistance (the opposition to the voltage), and current (the flow of electrons in the circuit). Ohm's Law states that the amount of current flowing in a circuit is equal to the applied voltage divided by the circuit resistance.

PERIOD - The time required to complete one cycle of AC and is calculated as the reciprocal of the frequency ($1/f$). It is measured in seconds and designated with the letter T.

PHOTOVOLTAIC EFFECT - The generation of an electrical current in a circuit containing a photo-sensitive device when the device is illuminated by visible or nonvisible light.

POWER - The rate of the time it takes for work to be done and is measured in watts (W). In electrical and electronic circuits, Power (P) = Supply Voltage (E) x Supply Current (I) or Load Voltage (V_L) x Load Current (I_L)). See WATT.

PROTECTED AREA - An area equipped with appropriate ESD protective materials and equipment. It provides a site where ESD voltage is limited below the electrostatic discharge sensitivity level of the component or equipment being handled or manufactured.

PULSATING DC VOLTAGE - Rectified AC voltage, either positive or negative, with respect to a reference. Half-wave pulsating DC voltage uses only one-half of the available AC voltage. Full-wave DC voltage uses both halves of the AC voltage waveshape.

REFERENCE - An arbitrarily selected point or section of a circuit or system to which the polarities and values of the circuit voltages are referred. See GROUND.

RELIABILITY - The assurance that a component will perform in a specified manner for a specified time under a set of specified conditions that include electrical, mechanical, thermal, and environmental stresses. The concept of reliability encompasses the elements of both quality and longevity. See STABILITY.

RESISTANCE - The electrical characteristic of a component, material, circuit, or system which acts to limit current in a circuit. It is measured in ohms (Ω) and designated with the letter R. Resistance depends on the molecular structure and dimensions of a component or device and on the configuration of a circuit or system. See OHM.

SINE WAVE - A smooth, continuously moving waveshape that has no break in its appearance. It has both positive and negative half-cycles generally symmetrical with respect to a reference. The cyclical repetition of these waves produces a waveshape that has a specified frequency in hertz (number of cycles per second) and a specified amplitude.

SQUARE WAVE - A rectangular-shaped periodic wave with a positive and negative half-cycle of equal times or widths. A square wave consists of a sine wave's fundamental frequency combined with the odd harmonics (multiples) of the fundamental frequency.

STABILITY - The ability of a component, circuit, or system to maintain a fixed level of operation, within specified tolerances, under varying external conditions. Changing conditions include voltage, frequency, temperature, and longevity. See RELIABILITY.

STEADY-STATE DC VOLTAGE - A fixed polarity of positive or negative voltage with respect to a reference. This form of voltage is used as the power source for electronic circuits.

TEMPERATURE COEFFICIENT (TC) - The change in the characteristic of a component which occurs because of a change in temperature. TC can be specified either as the number of parts per million (ppm) change in value per °C change in temperature, or as a percent change in value per °C change in temperature.

TRIBOELECTRIC EFFECT - The phenomenon of transferring electrons from one nonconductive material to another when friction is produced between them. See ELECTROSTATIC CHARGE.

TURBINE - A mechanical device with blades mounted onto its assembly and mechanically coupled to an electrical generator. When a turbine is placed in the path of flowing water, steam, or moving air, the movement of the water, steam, or air across the blades causes them to turn. The generator's armature rotates within a magnetic field which then produces electrical energy at the terminals of the generator. See ELECTRICAL GENERATOR.

VARISTOR - a metal (zinc) oxide over-voltage protective device. See METAL OXIDE ELEMENT.

VOLT - The unit of measurement of electromotive force necessary to produce one ampere of current in a circuit having a total resistance of one ohm. The volt is named for Alessandro Volta, an 18th Century Italian physicist.

VOLTAGE - The electromotive force that exists across a voltage source (supply voltage) or a load in a circuit. Its unit of measurement is a volt. See ELECTROMOTIVE FORCE.

VOLTAGE ARRESTOR - A fast-acting protective device that can absorb or short to ground a voltage in excess of the device's rated voltage.

WATT - The unit of measurement for electrical power, named after James Watt, an 18th Century Scottish engineer. One watt of power is dissipated when a voltage of one volt is applied across a load of one ohm resulting in one ampere of current in the circuit. See POWER.

WAVELENGTH - The physical distance between the beginning and the end of a cycle in a periodic wave (sine wave or square wave) as it travels through space or through a conductor. Wavelength is measured in meters (or occasionally in Ångström units) and is designated with the Greek letter lambda (λ).

GREEK LETTERS REPRESENTING ELECTRONIC AND MATHEMATICAL SYMBOLS

Greek letter	**Meaning**
β - beta	transistor current gain
Δ - delta	change
Ø - theta	angular displacement
λ - lambda	wavelength
μ - mu	micro (1 millionth or 10^{-6})
π - pi	mathematical constant = 3.14
Σ - sigma	mathematical sum
T - tau	period of time
Ω - omega	ohms

APPENDIX

ANSWERS TO REINFORCEMENT EXERCISE

Questions are listed on pages 101 - 103.

1. True - This is the relationship known as Ohm's Law.

2. True - Change in current is proportional to the change in voltage.

3. False - As the resistance in a circuit decreases with the supply voltage remaining constant, the current increases. According to Ohm's Law, $I = E/R$.

4. True

5. True - In a series circuit, the total or equivalent resistance of the individual loads in the circuit is the sum of the loads. Inserting additional resistance in series in the circuit increases the total or equivalent resistance, resulting in less current being drawn from the supply.

6. True - This is one of the rules of a series circuit.

7. False - In a parallel circuit, the addition of more loads connected in parallel reduces the equivalent resistance of the circuit, resulting in more current being drawn from the supply. With the addition of the parallel load, more current is being supplied from the voltage source which adds to the original current from the voltage supply.

8. True

9. True

10. False - Although copper is the material most commonly used as a conductor, silver has a lower resistance (higher conductivity) than copper.

11. True

12. True - Mechanical circuit breakers, however, can eventually wear out because of normal usage, metal fatigue, and friction.

13. True

14. False - Power is equal to the product of voltage (E or V) times current (E x I or V x I). If the supply voltage (E) is doubled, with the load resistance (R_L) remaining constant, the load current (I_L) will also double. If the current doubles and the voltage is doubled, the power dissipation is quadrupled.

15. True

16. False - In most circuit situations, "heat is the enemy" and must always be considered when designing a circuit.

17. False - As the frequency of a sine wave increases, the time required for one cycle (its period) and the distance between peaks of one cycle (its wavelength) decreases. The higher the frequency, the shorter both the period and wavelength.

18. False - Although the line voltage generated by the power companies throughout the country is in the form of AC, it is not suitable for use as the final source of power for electronic circuits used in home equipment such as radio, television, and audio systems.

19. True

20. True - Steady-state DC is the type of voltage that is used as the final voltage source for electronic circuits. A battery can be used to supply this form of voltage. If AC voltage is used as the prime source of electrical power, it must be converted to steady-state DC voltage for use in electronic equipment.

CALCULATION OF VOLTAGE, CURRENT, AND POWER

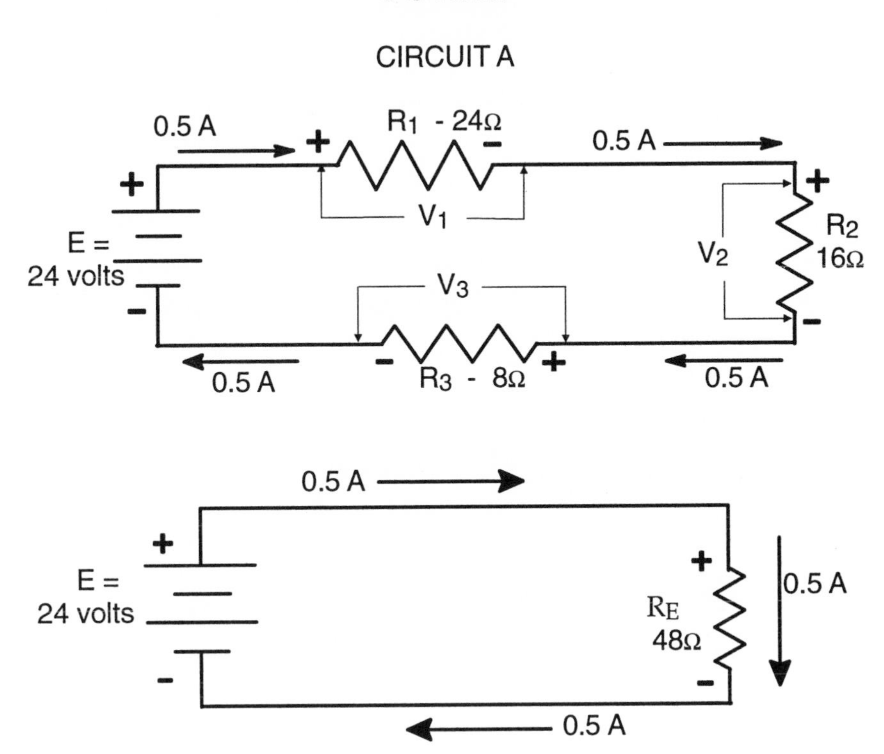

Since this is a series circuit, the equivalent resistance is the sum of each load resistance and is calculated as follows:

$$R_E = R_1 + R_2 + R_3$$
$$24 + 16 + 8 = 48 \text{ ohms}$$

With a 24 volt supply, the load current is calculated as follows:

$$I = \frac{E}{R} = \frac{24}{48} = 0.5 \text{ ampere}$$

As the 0.5 ampere load current leaves the positive terminal of the voltage supply in the original circuit , it produces:

12 volts across R_1 ($V_1 = I \times R_1$)
8 volts across R_2 ($V_2 = I \times R_2$)
4 volts across R_3 ($V_3 = I \times R_3$).

Knowing the voltage across and the current through each load, the power consumed can now be calculated as 6 watts, 4 watts and 2 watts respectively. Adding these quantities, the total power consumed in the circuit is equal to 12 watts.

As a double check, the supply voltage (24 volts) times the load current (0.5 ampere) is equal to a total power dissipation of 12 watts.

CIRCUIT B

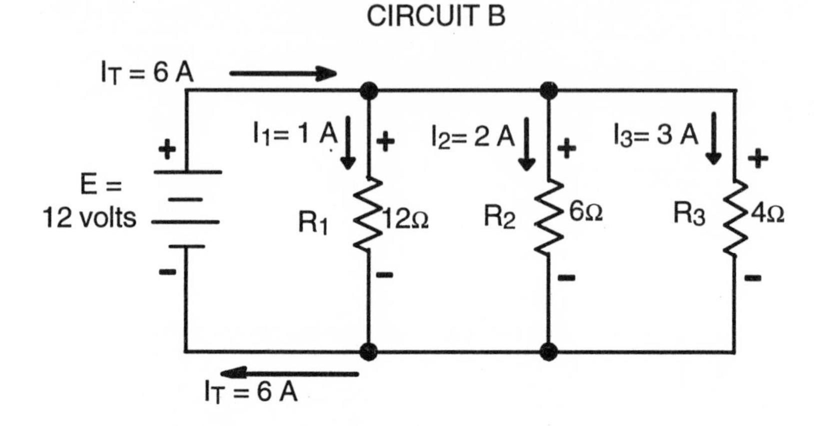

In a parallel circuit, each load has the same 12 volts supply voltage across it. Knowing the voltage and the values of each load resistance, the current through each load can be calculated. Using Ohm's Law ($I = V/R$), the load currents are calculated as:

$$V_1 = V_2 = V_3 = E = 12 \text{ volts}$$

$$I_1 = V_1/R_1 = 12/12 = 1 \text{ ampere}$$
$$I_2 = V_2/R_2 = 12/6 = 2 \text{ amperes}$$
$$I_3 = V_3/R_3 = 12/4 = 3 \text{ amperes}$$

Knowing the voltage across each load and the current through each load, the power consumed in each load is calculated as:

$$P_1 = V_1 \times I_1 = 12 \times 1 = 12 \text{ watts}$$
$$P_2 = V_2 \times I_2 = 12 \times 2 = 24 \text{ watts}$$
$$P_3 = V_3 \times I_3 = 12 \times 3 = 36 \text{ watts}$$
$$P_T = P_1 + P_2 + P_3 = 12 + 24 + 36 = 72 \text{ watts}$$

The same total power (P_T) can be calculated by multiplying the 12 volt supply voltage by the 6 ampere total load current (I_T). $P_T = E \times I_T$, (12 x 6) resulting in a product of 72 watts.

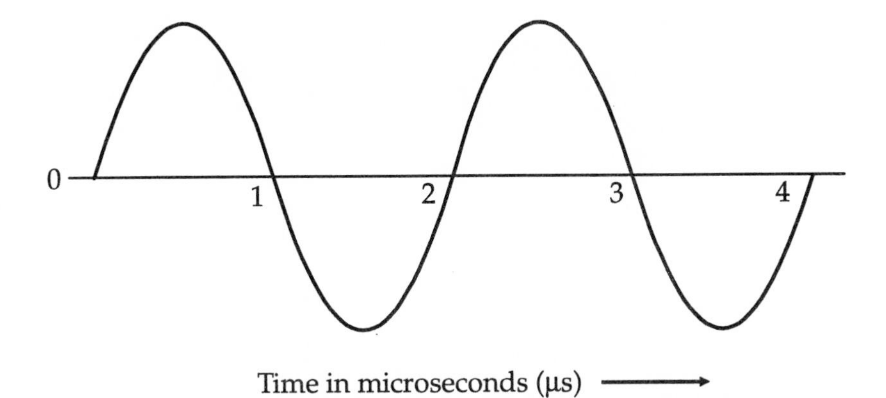

The period (T) of the sine wave is shown as 2 microseconds. The frequency (f) of a sine wave is equal to the reciprocal of the period:

Since 2 microseconds can be written as 2×10^{-6}, the frequency is calculated as follows:

$$\text{frequency} = f = \frac{1}{T}$$

$$f = \frac{1}{2 \times 10^{-6}} = \frac{1 \times 10^{6}}{2} \text{ Hz}$$

$$f = \frac{1}{2} \times 10^{6} = 0.5 \times 10^{6} \text{ Hz}$$

$$f = 0.5 \text{ megahertz (MHz), or}$$

$$f = 500 \times 10^{3} \text{ Hz} = 500 \text{ kilohertz (kHz)}$$

PART TWO

PASSIVE COMPONENTS

INTRODUCTION TO PASSIVE COMPONENTS

Components are devices that are interrelated in a specified manner to create circuits and systems. When connected to a source of voltage, they perform specific electrical or electronic functions and are primarily classified as either **passive components** or **active components**, depending on their electrical characteristics and capabilities.

Passive components are devices that can sense, monitor, transfer, attenuate, vary, or control voltage or current. They have the capability of functioning in other applications, such as voltage storage, switching, and timing. These components possess unique features that are necessary to make electronic circuits and systems function properly and play a significant role in circuit technology.

A passive component cannot differentiate between positive or negative voltage polarity nor can it, by itself, provide any gain or amplification in a circuit. These specific capabilities are the function of active components called **semiconductors**.

The subject of active components is covered in Volume Two - Part One - Discrete Semiconductors.

PASSIVE COMPONENT CATEGORIES

Fixed resistors
Variable resistors:
- Potentiometers
- Rheostats
- Trimmers

Capacitors
Mechanical switches
Keyboards and keypads
Electromechanical relays
Inductors, coils, and chokes
Transformers
Connectors
Indicator lamps
Crystals

Part Two of this volume provides the details of how passive components are made and packaged, how they work, how they are specified on a data sheet, and how they interrelate with other components in circuitry.

9

CHAPTER NINE

RESISTORS

FIXED RESISTORS

- FUNCTIONS AND SPECIFICATIONS
- FIXED RESISTOR TYPES
- COLOR CODING

VARIABLE RESISTORS

- POTENTIOMETERS
- RHEOSTATS
- TRIMMERS

CHIPS FOR HYBRID CIRCUITS

NETWORKS

ATTENUATOR CIRCUITS

- STEP ATTENUATOR
- VARIABLE ATTENUATOR

REINFORCEMENT EXERCISE

RESISTORS

Ohm's Law states that current (**I**) is equal to voltage (**E** or **V**) divided by resistance (R). Since resistance is one of the three basic parameters of Ohm's Law, it must be present in every circuit to allow that circuit to function.

A resistor is a standard component and is available as either:
- A fixed resistor having a specific value in ohms
- A variable resistor with an adjustable range of specified values

FIXED RESISTORS

FUNCTIONS

- A resistor can function as a circuit load. The current flowing in the circuit will depend on the voltage source, **E**, and the value of the load resistor($\mathbf{R_L}$).

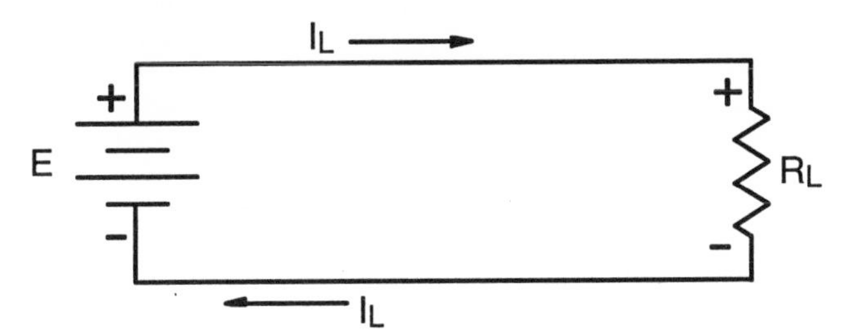

Using a Resistor as a Circuit Load
Figure 9.1

- A series resistor ($\mathbf{R_S}$) can be part of a circuit load in conjunction with another device ($\mathbf{R_{Lamp}}$) to limit the current in the circuit.

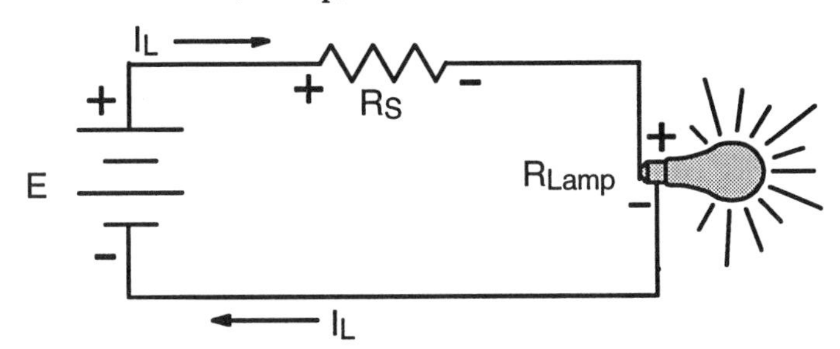

Using a Resistor as Part of a Circuit Load
Figure 9.2

- A resistor can produce a desired value of voltage in a circuit. For example, if a resistor is inserted in series in an existing circuit, a new voltage (**V_N**) is developed across the new resistor (**R_N**). This new voltage is the product of the new current (**I_N**) flowing in the circuit times the value of the new resistor:

$$V_N = I_N \times R_N.$$

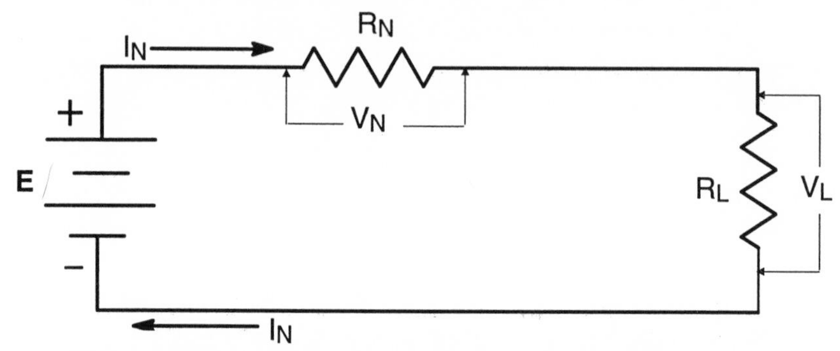

Adding a Resistor in Series to Obtain a Desired Voltage (V_N)
Figure 9.3

- A group of two or more resistors in series connected across a supply voltage can share the supply voltage in a ratio that depends upon the value of each resistor compared with the total resistance of all the resistors in series. This configuration is called a *resistor chain* or a *voltage divider*.

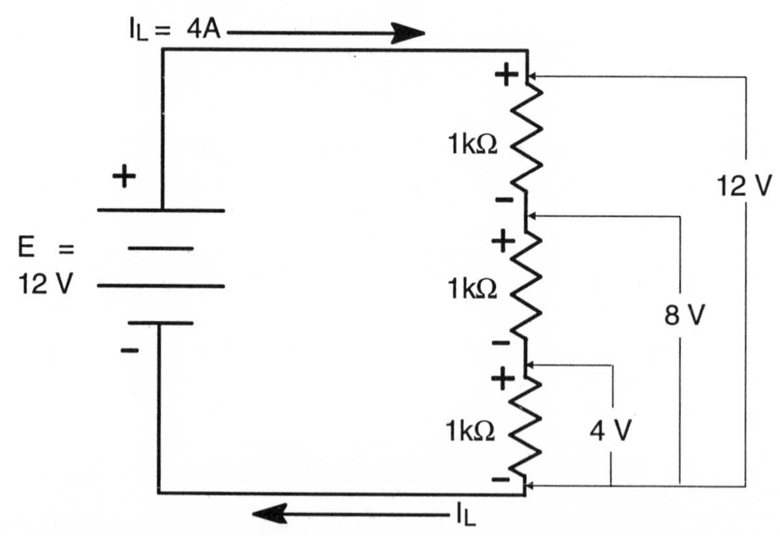

Resistor Network Connected as a Voltage Divider
Figure 9.4

- When connected in series or parallel with a capacitor, a resistor can function as part of a resistor-capacitor (RC) network, the main section of an electronic timing circuit. (See Figures 9.5 and 9.6)

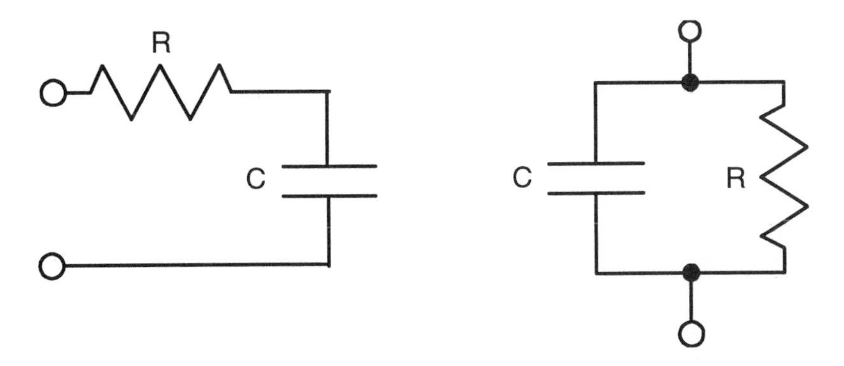

RC Networks for Electronic Timing Circuits
Figure 9.5 **Figure 9.6**

FIXED RESISTOR SPECIFICATIONS

When specifying a fixed resistor, its electrical, physical, thermal, environmental features, and limitations must be considered.

ELECTRICAL CHARACTERISTICS

RESISTANCE - The nominal value of the resistor specified at +25°C

TOLERANCE - Expressed as a plus or minus (±) percentage of the nominal value

MAXIMUM POWER CAPABILITY - Specified at +25°C, with power derating information for elevated temperatures

VOLTAGE RATING - The maximum DC or AC voltage allowed across a resistor without breakdown of the resistor material. Most resistors are capable of handling more than 1000 volts, as long as the power rating is not exceeded. (For most applications, the applied voltage is between 5 and 24 volts). Unless very high voltage is required across the resistor, its voltage rating is usually not specified.

ELECTRICAL NOISE - Most resistors create random, unwanted voltages called *noise*, caused by the molecular agitation of the material. Noise level depends on the type of resistor, its resistance value, the circuit configuration, and environmental conditions.

Noise can be specified as volts (microvolts) or as a decibel (db) value. The decibel specification is expressed as a ratio between the noise voltage generated by the resistor and a reference value of noise under controlled environmental conditions specified in the manufacturer's data sheet or customer's specifications.

PHYSICAL CHARACTERISTICS

The resistor's physical size, terminal lead material, lead length, and mounting features are usually described in a drawing on a manufacturer's data sheet or customer's specification drawing.

THERMAL CHARACTERISTICS

OPERATING AND STORAGE TEMPERATURE RANGE

- For military and space equipment:
  -55°C to +125°C (-67°F to +257°F)
- For industrial, commercial, and consumer applications:
  0°C to +70°C (+32°F to +158°F)

The specific temperature range can vary, depending on the individual application and is listed in the manufacturer's data sheet or the customer's specification sheet.

TEMPERATURE COEFFICIENT (TC) - This thermal characteristic of a resistor produces a change (ΔR) in its nominal value (R at +25°C) when subjected to a change in temperature (ΔT).

- Positive TC (PTC)
  Resistance **increases** with **increasing** temperature

- Negative TC (NTC) -
  Resistance **decreases** with **increasing** temperature

Temperature coefficient (TC) can be expressed in **parts per million** (ppm) change in resistance for every °C change in temperature.

If a 5600Ω resistor with a TC of 1000 ppm per °C is subjected to a 10°C temperature change, its value would change by 56Ω. This change in resistance (ΔR) is calculated by multiplying the value of the resistor at +25°C by the **TC** and then multiplying by **ΔT**.

Calculation of resistance change (ΔR) → $\mathbf{\Delta R = R \times TC \times \Delta T}$

$$\Delta \mathbf{R} = 5600 \times \frac{1000}{1{,}000{,}000} \times 10 = 5600 \times .01 = 56\Omega$$

TC can also be expressed as a **percent change in resistance** for every °C change in temperature.

- In using this second technique of expressing TC, the 1000 ppm is converted to 0.1% change in resistance. The change in resistance (Δ**R**) is calculated as before by multiplying the value of the resistor at +25°C by the **TC** and then multiplying by the change in temperature (ΔT).

  $\Delta\mathbf{R} = 5600 \times 0.1\% \times 10 = 5600 \times .001 \times 10 = 5600 \times .01 = 56\Omega$

Whichever method is preferred, the chart for converting one method into another for different values of TC is shown in Figure 9.7.

PPM ΔR/°C TO % ΔR/°C			
1 ppm .0001%	10 ppm .001%	25 ppm .0025%	50 ppm .005%
100 ppm .01%	200 ppm .02%	500 ppm .05%	1000 ppm 0.1%

Conversion from ppm ΔR /°C to % ΔR/°C
Figure 9.7

ENVIRONMENTAL CONSIDERATIONS
The resistor's ability to withstand environmental stresses, such as: vibration, mechanical shock, humidity, fungus, salt spray, chemical solvents, etc.

FIXED RESISTOR TYPES

To optimize selection for specific fixed resistor application, choices can be made from a variety of materials. These include:

- Carbon composition
- Carbon film (also called deposited carbon)
- Metal film
- Wirewound (nichrome or other resistance wire)
- Glazed metal
- Cermet film (combination of ceramic and metal materials)
- Metal oxide film
- Bulk metal film
- Thermistors
- Varistors

CARBON COMPOSITION

Carbon composition resistors are used for general purpose applications. A filler and binder are mixed with carbon powder and then formed into a cylinder with leads anchored to its ends.

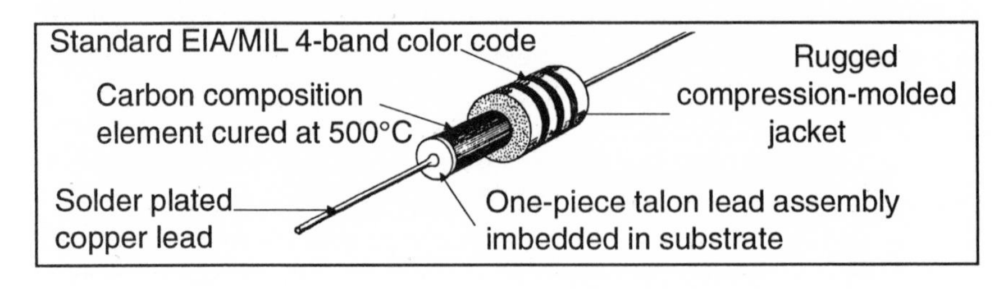

RANGE AND TOLERANCES
Carbon composition resistors range from 1 ohm to 22 megohms with tolerances of:
± 20%, ± 10%, and ± 5%.

Base values indicate the first two numbers of the resistance value, e.g.: a 3000 ohm resistor (base value = 30) is only available as a ± 5% tolerance type.
(See Figure 9.9)

The base values are established by the Electronics Industry Association (E.I.A.)

POWER CAPABILITY
Power capability is dependent upon the physical size of the resistor and ranges from ⅛ watt to 2 watts.

TC
The temperature coefficient of carbon composition resistors is relatively very high, about 6500 ppm/°C, or 0.65% change in resistance per °C.

± 20%	± 10%	±5%
10	10	10
		11
	12	12
		13
15	15	15
		16
	18	18
		20
22	22	22
		24
	27	27
		30
33	33	33
		36
	39	39
		43
47	47	47
		51
	56	56
		62
68	68	68
		75
	82	82
		91

Standard Base Values
Figure 9.9

Carbon composition resistors are intended for applications where there are no stringent requirements for resistance tolerance, temperature stability, or low noise. They are specified in commercial, industrial, or consumer equipment, and for military and space applications. These resistor types are sometimes used where instantaneous surge currents are present since they are capable of safely handling sudden overloads of current and power for a short time.

CARBON FILM or DEPOSITED CARBON

Specified for general purpose applications, they are approximately the same physical size as carbon composition resistors and are constructed by depositing carbon in film form onto a ceramic core. (See Figure 9.10) Leads are attached to the ends of the body and a spiral is cut through the resistive coating to lengthen the resistive path and, thereby, adjust its resistance. The resistor is then coated with an epoxy material conforming to its body shape. This technique of coating is called *conformal-coating*.

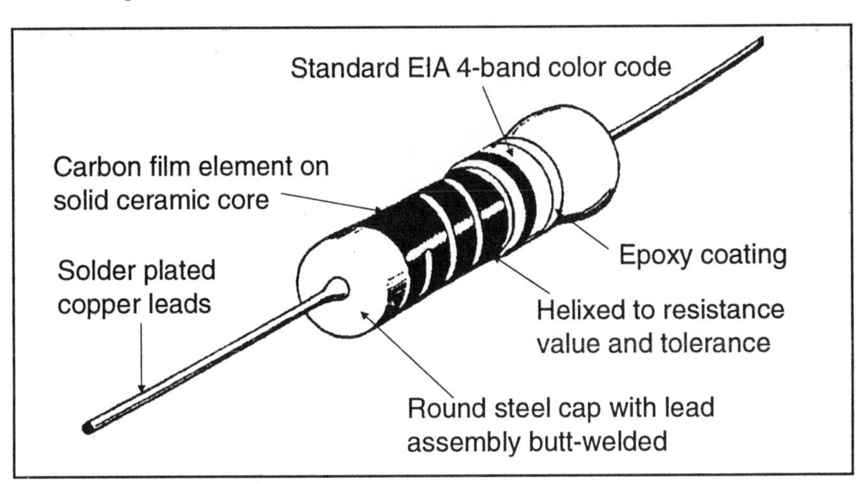

Carbon Film Resistor Construction
Figure 9.10

Carbon film resistors are manufactured with resistance values and power ratings similar to those of carbon composition types (1 ohm to 22 megohms and ⅛ watt to 2 watts). Tolerances range from ± 2% to ± 10%, with the TC about one third that of carbon composition (approximately 2500 ppm/°C). Carbon film resistors, however, tend to generate more electrical noise than carbon composition resistors. The epoxy coating provides physical protection and resistance to certain environmental stresses.

METAL FILM

These resistors are used in circuits where high precision and low temperature coefficient characteristics are required.

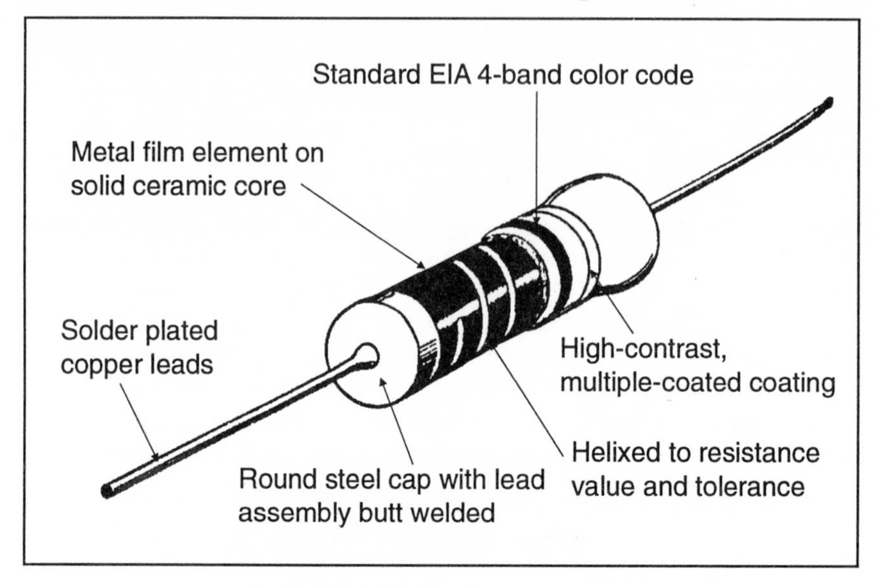

Metal Film Resistor Construction
Figure 9.11

These devices are made by depositing a resistive metal alloy, in film form, onto a ceramic or glass core. After leads are attached and the spiral has been cut, the body is encapsulated in an epoxy conformal-coating or a molded plastic jacket for protection against environmental stresses.

Metal film resistance values range from 0.27 ohms to 10 megohms, with tolerances ranging from ± 0.1% to ± 10%. The TC of metal film resistors is about one-tenth that of carbon film (from 10 to 250 ppm/°C), with power ratings ranging from 1⁄20 of a watt to 3 watts.

WIREWOUND

Used for both high power or ultra-precision resistors, they are made by winding high-resistance metal alloy wire (usually nichrome, a combination of nickel and chromium) around a ceramic core. The body is then coated with vitreous enamel, silicone, or a cement compound. These materials provide physical protection, are fire-retardant, and are resistant to the environmental stresses of humidity, salt spray, fungus, and chemical solvents.

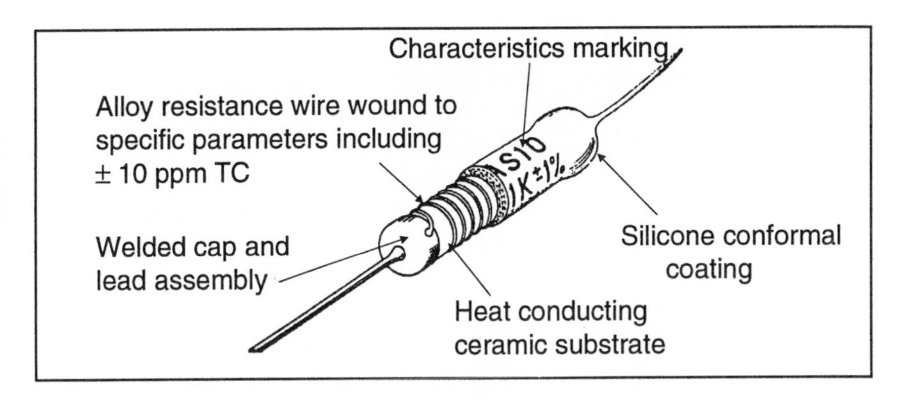

Wirewound Resistor Construction
Figure 9.12

Wirewound resistance values range from several thousandths of an ohm to about 1 megohm, with tolerances for the power types (3 watts and above) from ±1% to ±10%. Power capability of these types range from 3 watts to several hundred watts with very low temperature coefficients that range from 5 to 50 ppm/°C.

Precision and ultra-precision wirewound resistors are also available with the same material and construction, having power capabilities of 2 watts or less, with temperature coefficients as low as 5 ppm and tolerances as low as ± .01%.

Not considering temperature effects, high resistance wire offers a constant resistance to a DC voltage. In addition to this DC resistance, an *inductance* is created. Inductance produces resistance to an AC voltage and is an inherent characteristic of a wire that is wound around a core. This AC resistance is called *inductive reactance* and is measured in ohms.

As the frequency of an applied AC voltage changes, the inductive reactance of a wirewound resistor changes accordingly, but its resistance to DC remains the same. Other resistors, not wirewound types, oppose both DC and AC equally despite the changing frequency of an applied AC voltage.

The inductive reactance of an ordinary wirewound resistor makes the device undesirable when used with varying frequency AC. For high power AC, *non-inductive* resistors are wound in a manner that reduces the effect of inductance and provides the desirable features of wirewound resistors without the undesirable change in AC resistance as the frequency of the applied voltage changes.

A more detailed discussion of inductance and inductive reactance is covered in Chapter Twelve - Magnetic Components.

GLAZED METAL and CERMET FILM

These devices are used for very high resistance requirements and are exceptionally effective in withstanding extreme environmental stresses. They are made by depositing or screening a glaze of metal alloy onto a ceramic rod. The resistor is fired in an oven at a very high temperature and then enclosed in a molded plastic jacket or given an epoxy conformal-coating. (See Figure 9.13)

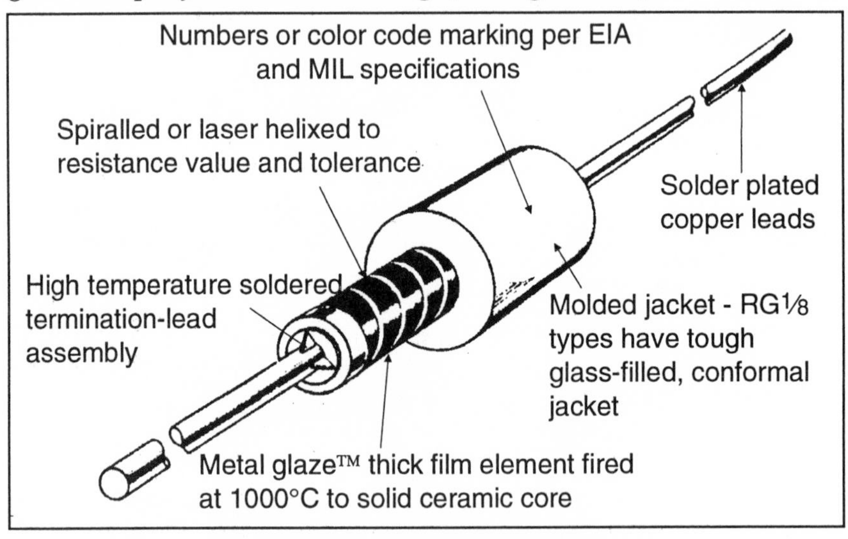

Glazed Metal Resistor Construction
Figure 9.13

The cermet types are available with extremely high resistance, up to 50 gigohms in value. Tolerance values range from ± 1% to ± 5% with a TC of about 10 ppm/°C.

METAL OXIDE FILM

These resistors are used where very high resistance, low electrical noise, and high stability characteristics are required. They are made by spraying or vaporizing a tin-chloride solution on the surface of a heated glass substrate. Additives are used on the resulting tin-oxide film to produce the desired characteristics. They are capable of withstanding high power overloads for long pulse durations. Temperature coefficients range from 50 to 250 ppm/°C.

BULK METAL ELEMENT (FILM)

Designed for high frequency and high speed applications, these devices are essentially non-inductive. Their characteristics are similar to precision wirewounds and are virtually noise-free. The temperature coefficients range from 5 to 50 ppm/°C.

SUMMARY OF FIXED RESISTOR CHARACTERISTICS

Resistor Type	Resistance Range (Ohms)	Tolerance Range (± %)	Power Range (Watts)	Temperature Coefficient Range (PPM/°C)
CARBON COMPOSITION	10 to 22M(1) 1 to 100M(2)	5 to 20	⅛ to 2	about -6500
CARBON FILM	1 to 22M(1) 1 to 200M(3)	2 to 10	⅛ to 2	-800 to +2500
METAL FILM	0.27 to 10M	0.1 to 10	1/20 to 3	± 10 to ± 250
WIREWOUND Power	0.1 to 1M	1 to 10	3 to 1500	± 5 to ± 50
Precision	0.1 to 1M	.001 to 1	.05 to 2	± 1 to ± 50
Ultra Precision	0.1 to 1M	.01 to 0.1	.05 to 2	± 1 to ±10
GLAZED METAL	1 to 50G	1 to 5	0.1 to 10	about ± 10
METAL OXIDE	10 to 15M(4)	1 to 5	¼ to 115(5)	± 50 to ± 250
CERMET FILM	10 to 50G	1 to 5	0.1 to 2	about ± 10
BULK METAL ELEMENT (FILM)	30 to 10M	.01 to 1	.05 to 0.75	± 5 to ± 50

(1) Standard
(2) Hot-molded type - to 1 teraohms in specials
(3) To 100 teraohms in specials
(4) To 1500 MW in high-voltage types and 1 teraohms in specials
(5) To 400 watts in specials

Figure 9.14

THERMISTORS

The word "thermistor" is derived from "thermal resistor" since it is a resistor that reacts to changes in temperature in a circuit. It has a precisely specified temperature characteristic to provide controlled changes in a component's ohmic value as its temperature changes.

POSITIVE TEMPERATURE COEFFICIENT (PTC) THERMISTORS

- PTC thermistors are made from powders containing barium titanate, strontium, or lead, and traces of rare earth elements.

- PTC types are relatively low in resistance initially, and they remain at a constant value with a small increase in temperature. At the critical switching temperature, the resistance value increases very suddenly. Over a 100°C range, starting at the switching temperature, the increase in resistance can be as much as seven hundred times more than its initial low value.

- The characteristic of the thermistor makes the device behave like an ideal switch that can be controlled by a change in temperature.

- The following applications are made possible: resettable and time-delay fusing, protection against surge or in-rush current, motor starting, and combination heater-thermostats.

NEGATIVE TEMPERATURE COEFFICIENT (NTC) THERMISTORS

- Generally made from a combination of nickel oxide, cobalt oxide, and manganese oxide; carbon powders are often added to this mixture. As the temperature increases in an NTC thermistor, its resistance decreases.

- NTC characteristic causes the thermistor to function as a temperature compensating device to offset PTC changes of other components to provide temperature stablity in critical sections of a circuit. The circuit will then operate with no change in its parameters, regardless of changing temperature. This is considered to be a *zero TC* operating condition.

- Other applications for an NTC thermistor include temperature control and temperature measurement circuits. The circuit shown in Figure 9.15 illustrates the use of an NTC thermistor in conjunction with an appropriately calibrated ammeter to measure the temperature of the surrounding atmosphere.

When the temperature increases, the resistance of the NTC thermistor decreases, allowing more current to flow in the meter, indicating a reading of increasing temperature. A decrease in temperature will produce an opposite effect on the thermistor, causing less current to flow, thereby reducing the meter readout.

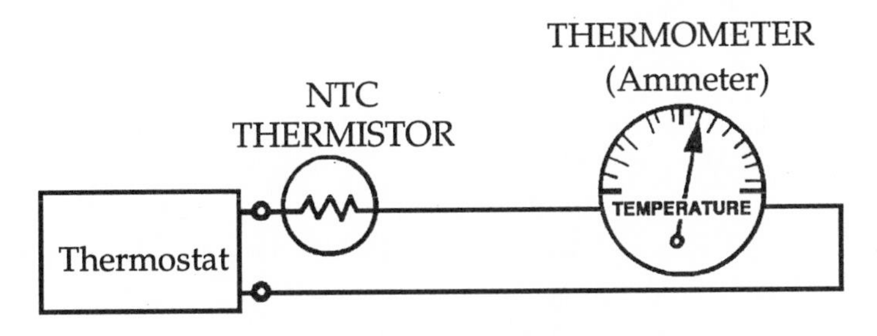

Thermometer Circuit Using NTC Thermistor
Figure 9.15

THERMISTOR PACKAGES

Thermistors are manufactured in a wide variety of packages - small and large disks, beads, rods, and chips. When used as a temperature probe, they are available in a multitude of different styles to suit any temperature probe application.

VARISTOR

This metal (zinc) oxide device has variable resistor characteristics that act to protect circuitry against lightning and other excessive voltage surges. Resistance varies as a function of applied voltage.

Below its rated voltage (95 to 1000 volts), its extremely high resistance has no effect on the circuit. Above its rated voltage it suddenly changes to an extremely low resistance, acting to clamp the applied voltage to a safe level of operation.

COLOR CODING

If a resistor is not marked with its numerical resistance value on its surface, the standard Electronic Industries Association (EIA) color code is used. Sequential color bands are painted on one end of the resistor body to designate the resistor's value and tolerance.

For commercial, industrial, and consumer types, the four-band EIA color code is used. Starting at the color band end:

- The first two bands indicate the first two significant digits of its nominal ohmic value.
- The third band indicates the number of zeros after the first two numbers.
- The fourth band indicates the tolerance of the resistor.

For high reliability, military type resistors, a fifth band is used to indicate the *established reliability* (ER) rating, which identifies the percent failure rate per thousand hours of operation of the resistor.

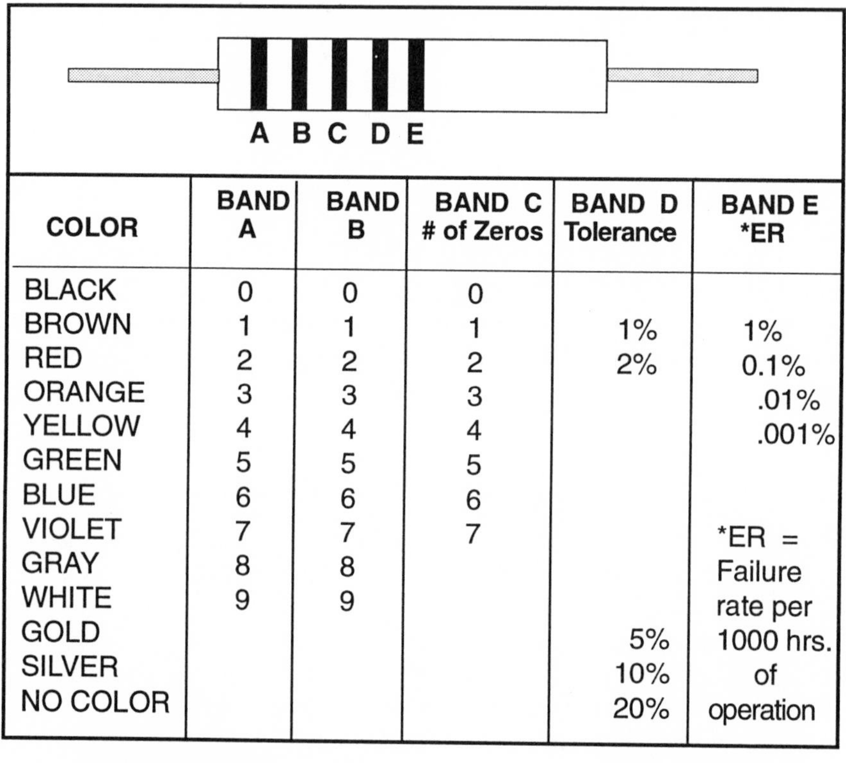

COLOR	BAND A	BAND B	BAND C # of Zeros	BAND D Tolerance	BAND E *ER
BLACK	0	0	0		
BROWN	1	1	1	1%	1%
RED	2	2	2	2%	0.1%
ORANGE	3	3	3		.01%
YELLOW	4	4	4		.001%
GREEN	5	5	5		
BLUE	6	6	6		
VIOLET	7	7	7		*ER =
GRAY	8	8			Failure
WHITE	9	9			rate per
GOLD				5%	1000 hrs.
SILVER				10%	of
NO COLOR				20%	operation

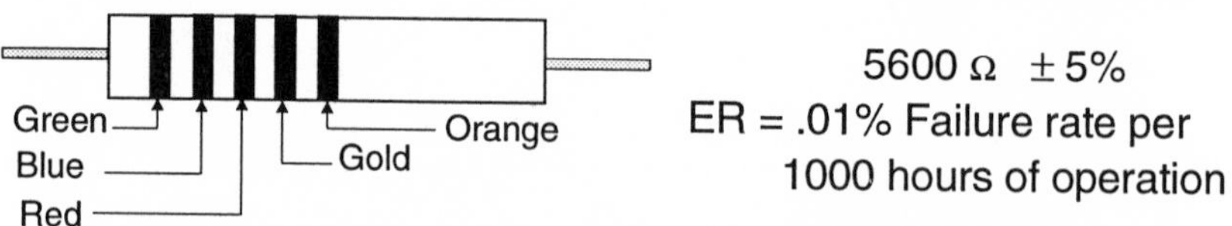

Resistor Color Code and Color Coding Example
Figure 9.16

In Figure 9.16, the first band is green, indicating number 5 and the second band blue, indicating number 6, making 56 the first two significant digits. The third band is red, indicating 2 zeros after the first two numbers. The resistance value is 5600 ohms or 5.6 kilohms. The fourth band is gold, indicating a tolerance of ± 5%.

In this example, there is a fifth band (orange), showing a MIL-type resistor with an established reliability rating of 0.01% failure rate per 1000 hours of operation.

VARIABLE RESISTORS

Variable resistors are used for voltage, resistance, and current adjustment. They cover a wide area of applications: volume and tone control in audio systems, timing, light and heat control, motor speed control, frequency tuning, circuit trimming, and calibration.

POTENTIOMETERS

A **potentiometer** is a three-terminal resistive device shown graphically in Figure 9.17; it is normally connected in a circuit with a voltage across its entire resistance, A-C. A portion of this voltage can be monitored, or metered, by tapping off the voltage that exists between the adjustable (control) terminal, B, and either end terminal. The voltage, E, across terminals A and C will not change, but the voltage, V, between the control terminal B and terminal C will vary, depending upon the control terminal setting. (See Figure 9.18)

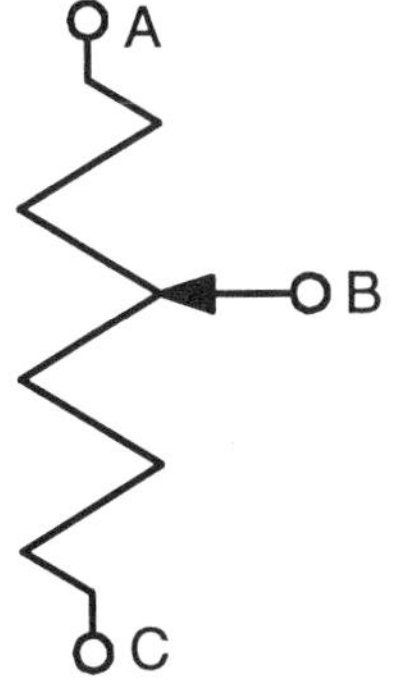

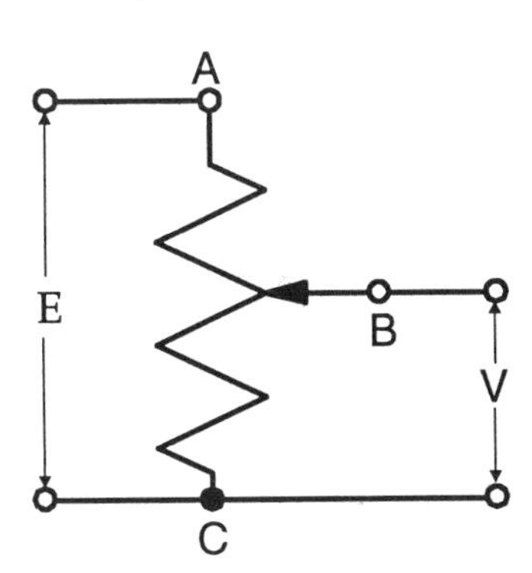

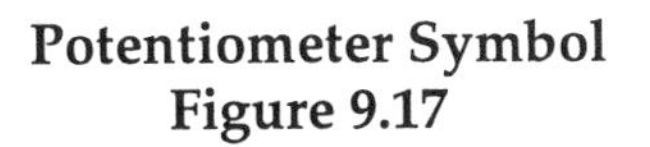

Potentiometer Symbol
Figure 9.17

Voltage Adjustment
Figure 9.18

The nearer point B is to point A, the higher the value of voltage (V). Conversely, as the control terminal is moved nearer to point C, the voltage, V, will decrease.

The setting variation is achieved by rotation of a shaft, movement of a slide control, or by a screw adjustment. Although the potentiometer is classified as a variable resistor, it is actually a variable voltage control device. As the control setting is changed, the voltage across the control terminal and either end terminal is changed; the resistance between the two end terminals remains the same.

RHEOSTATS

When only two of the three terminals are used (the control terminal and either end terminal) it is truly a variable resistor called a **rheostat**. Adjustment of the control produces a change in resistance between its two terminals. A rheostat is shown with two different graphic symbols in Figure 9.19).

Rheostat Graphic Symbols
Figure 9.19

The circuit in Figure 9.20 shows the use of a rheostat to adjust current. As the rheostat, R, is varied, the current, I, will vary inversely.

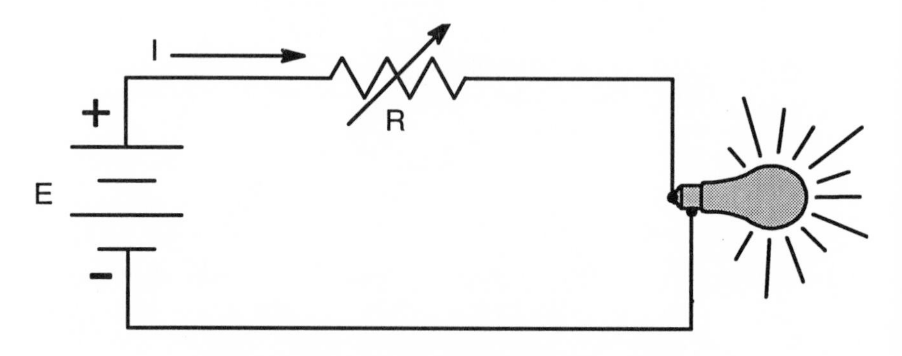

Using a Rheostat to Adjust Circuit Current
Figure 9.20

The terms "potentiometer" and "rheostat" refer to variable resistors that are externally and internally accessible and are used for controlling voltage or current levels. They are used as audio volume and tone controls, light-dimming, motor speed control, and heat control. The selection of either a potentiometer or a rheosat will depend on the desired circuit application.

TRIMMERS

A potentiometer or rheostat used for internal circuit adjustment or circuit calibration is referred to as a *trimming potentiometer*, *trimming rheostat*, or, generally, *trimmer*. A trimmer is designed to be adjusted with a screwdriver and, generally, does not accept a knob.

Although a trimmer is used for current control, voltage division, or timing control in the same way that a potentiometer or a rheostat is used, a trimmer generally works in conjunction with another fixed resistor or potentiometer to act as a fine adjustment control.

Trimmers can be mounted onto a bracket inside a system, inserted into a printed circuit board (PCB), or installed anywhere in the equipment so that the control can be either accessible or inaccessible to the end user.

Typically, trimmers are used in the maintenance and adjustment of a system, for circuit trimming, and calibration. These devices are often referred to as *set and forget* controls.

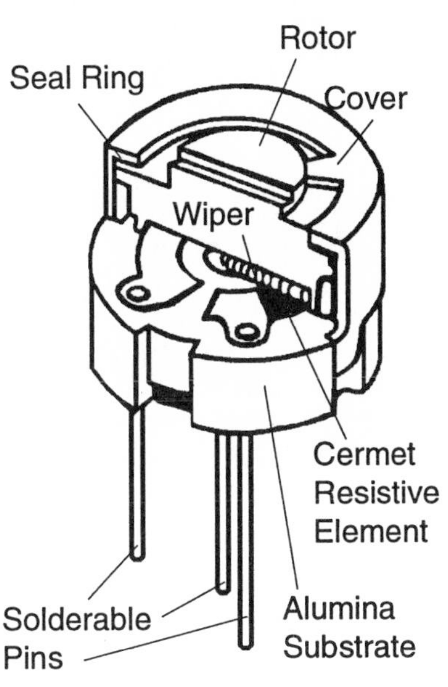

A Typical Single-turn Trimmer Construction
Figure 9.21

Normally, the function of a trimmer is to bring the total resistance of a section of a circuit to a precisely correct value. As an example, in the circuit of Figure 9.22, the 500 ohm trimmer is used to adjust the resistance between points A and B to exactly 1285 ohms. The trimmer provides an efficient way of maintaining this resistance to overcome the effects of associated components changing in value as they age.

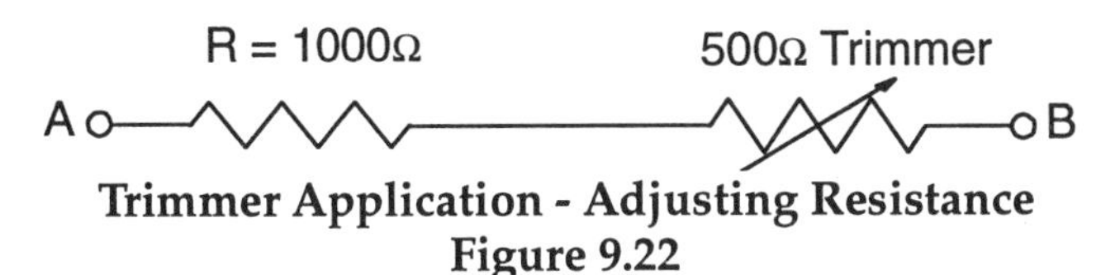

Trimmer Application - Adjusting Resistance
Figure 9.22

In Figure 9.23, a trimmer is used to provide a fine adjustment for the desired value of V_{OUT}, the voltage between terminals B and C.

With the proper choice of R_1, R_2, and R_3, the voltage range of V_{OUT} can be precisely set to the desired minimum/maximum values.

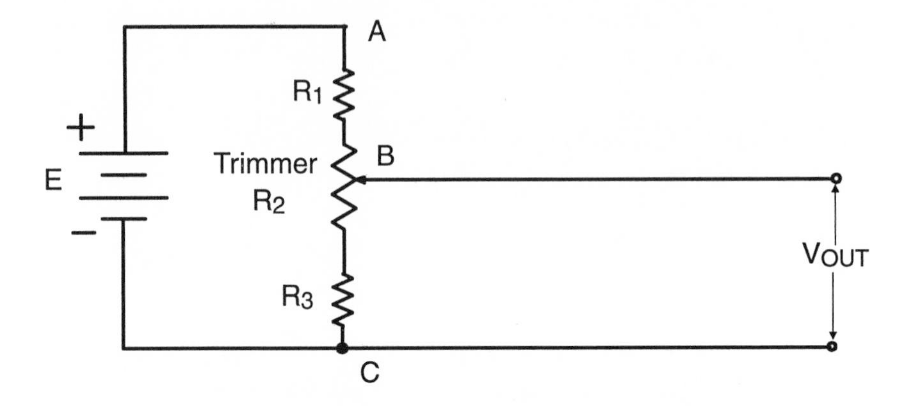

Trimmer Application - Adjusting Voltage
Figure 9.23

VARIABLE RESISTOR SPECIFICATIONS

Potentiometers, rheostats, and trimmers are available in a wide variety of materials, resistor values, styles, and mounting arrangements. As with fixed resistors, they are specified according to:

Ohmic value at 25°C
Tolerance
Power capability at 25°C
Temperature coefficient (TC)
Operating temperature and storage temperature range in °C
Ability to withstand environmental stress
Physical characteristics and dimensions
Mounting features (if applicable)

The additional characteristics, unique to variable resistors, include specifications for the *taper* of the control and the *number of turns* (single-turn or multi-turn) in a potentiometer, rheostat, or trimmer.

TAPER

The taper of a variable resistor indicates the characteristic of the resistance change as the control shaft is rotated.

Standard resistance taper curves (resistance vs. degree of rotation) are shown in Figure 9.24. In addition to these standard tapers, special tapers are available from most manufacturers of these components.

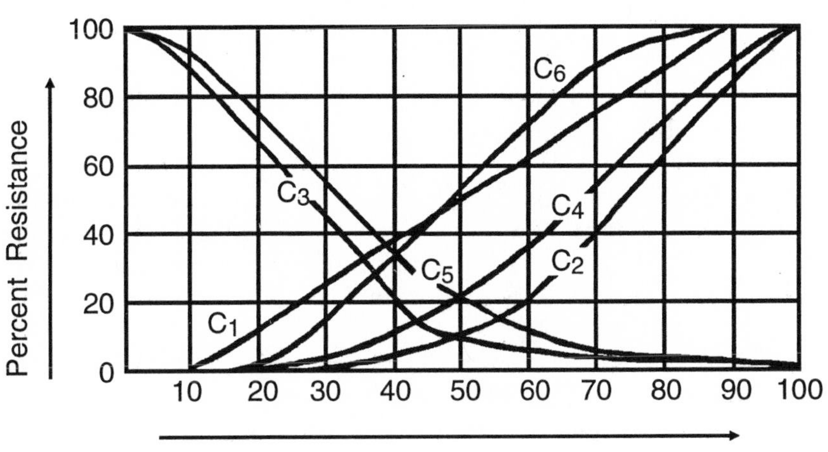

Standard Resistance Taper Curves
Resistance vs. Degree of Rotation
Figure 9.24

C_1 - Clockwise linear taper - for general purpose and television picture adjustments. Resistance is proportional to shaft rotation.

C_2 - Clockwise semilog taper - for volume and tone controls with 10% of its total resistance occurring at 50% of its shaft rotation.

C_3 - Counter clockwise semilog taper - reverse of C_2 - 90% of its total resistance occurring at 50% of its shaft rotation.

C_4 - Modified clockwise semilog taper - for volume and tone control use. 20% of its total resistance occurring at 50% of its shaft rotation.

C_5 - Modified counter clockwise semilog taper - reverse of C_4 - 80% of its total resistance occurring at 50% of its shaft rotation.

C_6 - Clockwise symmetrical taper with slow resistance change at either end - used as tone control or audio balance control.

NUMBER OF TURNS

In a single-turn variable resistor, the control contact travels the length of the resistance element in one revolution of the control shaft, or less. A single, complete rotation of the shaft is generally about 275°.

A multi-turn potentiometer, rheostat, or trimmer is one in which the control shaft is turned more than one revolution to allow the contact to traverse the complete length of the resistive element. Multi-turn controls are used for high precision resistance control and voltage adjustment.

For each 360° revolution of the control shaft, the control contact will traverse only a portion of the entire resistance element and the portion traversed will depend on the total number of turns in the device. For example, one revolution of a 3-turn potentiometer will control only one-third of the voltage applied across the potentiometer. Similarly, one revolution of a linear 3-turn rheostat will vary its resistance by ⅓ of the total value.

Multi-turn controls range from 3 to 40 turns with linearity as precise as ± 0.01%. When a multi-turn variable resistor (potentiometer or rheostat) is used as a front-panel control, a readout dial, called a *vernier*, is mounted on the shaft of the control so that it will be accessible on the front panel. The vernier is used to provide accurate calibration and fine adjustment of a variable resistor and permits precise control of its available accuracy and resolution.

> **Resolution** is the measure of the smallest possible incremental change in either the resistance of a rheostat, or the voltage at the control terminal of a potentiometer. The greater the number of turns of the shaft, the easier to achieve a small incremental change.

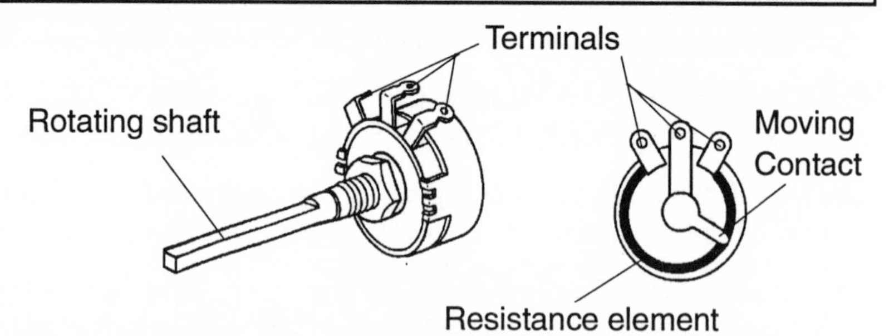

Typical Single-turn Potentiometer Construction
Figure 9.25

TYPES OF VARIABLE RESISTORS

CARBON COMPOSITION

Standard components range from 50 ohms to 5 megohms with tolerances of ± 5%, ± 10%, and ± 20%; they are made for general purpose applications where temperature and humidity are not critical. Custom potentiometers and rheostats are manufactured in values up to 1000 megohms. Power ratings range from ¼ watt to 3 watts and have relatively high TC. They are manufactured as single-turn controls only and are available as potentiometers and rheostats.

CARBON FILM

Used for general purpose applications and feature low noise and negligible frequency error, ranging from low frequencies to beyond 100 megahertz. Standard values range from 100 ohms to 10 megohms with tolerances of ± 10%, ± 20% and ± 30% and power ratings from ¼ to 1 watt. They have about one-third the TC of carbon composition devices and are manufactured as both potentiometers and rheostats.

METAL FILM and CERMET

These resistors are used where low TC, high stability, and resistance to environmental stress is required. Resistance values range from 50 ohms to 5 megohms with tolerances of ± 10% and ± 20%. Power capabilities range from ½ watt to 2 watts with a TC about .015% change in resistance per °C change in temperature. They are made as single-turn and multi-turn types for trimmer applications.

CONDUCTIVE PLASTIC

Used where low TC and small physical size are required, these resistors feature standard resistance values ranging from 100 ohms to 1 megohm in tolerances of ± 10% and 20%. These components are available as trimmers with power ratings of ¼ watt and ½ watt.

WIREWOUND

Wirewound controls are available as potentiometers, rheostats, or trimmers in a variety of packages, configurations, and control mechanisms. They are used for power (3 watts to 350 watts) and for precision applications (¼ watt to 2 watts). Resistance values range from 1 ohm to 250 kilohms with tolerances of ± 5% and ± 10%.

Because the moving contact slides from wire to wire, the resistance will vary in steps that may cause problems in some applications.

Just as certain effects result from the nature of a fixed wirewound resistor, the winding of high resistance wire around a core of a variable resistor produces an *inductance*. When an AC voltage of varying frequency is applied to a wirewound variable resistor, the changing frequency causes a change in its *inductive reactance*, (AC resistance). In some circuit applications, a variable wirewound resistor is undesireable because of this changing AC resistance.

The subject of inductance, inductive reactance and its effects in an AC circuit is covered in Chapter Twelve - Magnetic Components.

RESISTOR CHIPS FOR HYBRID CIRCUITS

A **hybrid circuit** consists of many components in chip form, assembled and interconnected on a pre-wired ceramic or glass substrate to form a circuit, network, subsystem, or array.

The component chips are soldered to the substrate or attached by thermal-compression bonding. Depending on the package design and end use, the external terminals are either solder plated or gold plated. Chips used for the resistor elements in these circuits are manufactured by either thick film or thin film technology.

THICK FILM TECHNOLOGY

Thick film resistor elements are made by a screen-printing process to produce a small resistive layer made of cermet, a combination of ceramic and various metals. The metals most often used are ruthenium dioxide and palladium-silver. The cermet is then fused to the substrate under high temperature.

Resistance values range from 10 ohms to 10 megohms with tolerances greater than ± 2%. If laser-trimmed, tolerances can be reduced to ±1%. The TC is greater than 200 ppm/°C with temperature tracking of adjacent thick film chips within ± 4%.

Resistor chips made with this process are used for hybrid integrated circuits where tolerance and temperature effects are not critical, e.g. commercial, industrial, and consumer applications.

THIN FILM TECHNOLOGY

Resistor chips are made by vapor-deposition of a very thin metallic film onto a ceramic or glass substrate. The metallic film is either nichrome, tantalum, or chromium-cobalt. Gold-plated leads are usually used as the interconnecting conductors on the substrate.

Resistance values for these chips range from 50 ohms to 100 kilohms with tolerances less than ± 1%. Typically, the temperature coefficient for thin film chips is 5 ppm/°C, with temperature tracking about ten times tighter than that of thick film. Thin film technology provides the characteristics needed for military and space applications - low temperature coefficient, tight tolerance, a wide operating temperature range, and a high degree of reliability.

Both thick film and thin film resistor chips are used in *Surface Mounted Device* (SMD) technology and in *resistor networks*, providing advantages in component assembly and space utilization. The subject of SMD technology is covered in Chapter Fourteen - Technology Trends For Passive Components.

RESISTOR NETWORKS

A **resistor network** consists of a group of individual resistors in chip form which can be arranged in a multitude of connection options in a single package. The resistors can be interconnected internally, providing various types of circuit configurations, or can be configured as individual, unconnected resistors.

Appropriate internal contact points are connected internally to the external terminals of the package. Options for packaging include the dual in-line package (DIP), single in-line package (SIP), flat-pack, and the "T" (subminiature) package. (See Figures 9.25a and 9.25b)

Resistor networks are also manufactured in a conformally-coated epoxy package for consumer applications. The package material selected depends on its end use and the temperature and environmental stresses to which the network will be subjected. For military and space applications, a Kovar or ceramic package with a hermetic (airtight) seal is used. For commercial and industrial applications where less severe environmental conditions exist, a less expensive plastic package is used.

PACKAGE OPTIONS

- Dual In-Line Package (DIP) - 14, 16, and 20 Pin
- Single-In-Line Package (SIP) - 6, 7, 8, and 10 Pin

Epoxy-molded jacket for ease in auto-insertion handling

60-40 soldered lead wires on 100 mil centers

Thick film cermet element fired at 1000°C, laser-trimmed to desired value and tolerance

2%

DIP

Resistive elements placed on the bottom for better heat dissipation

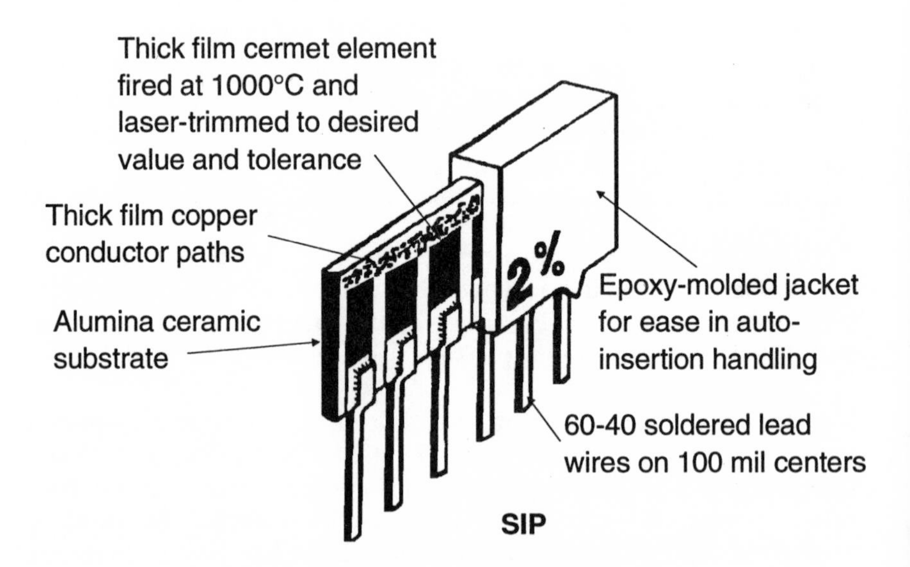

Resistor Network Package Options
Figure 9.25a

- Flat-Pac (FP) - 14 Pin
- Subminiature T-Package - 3 Pin

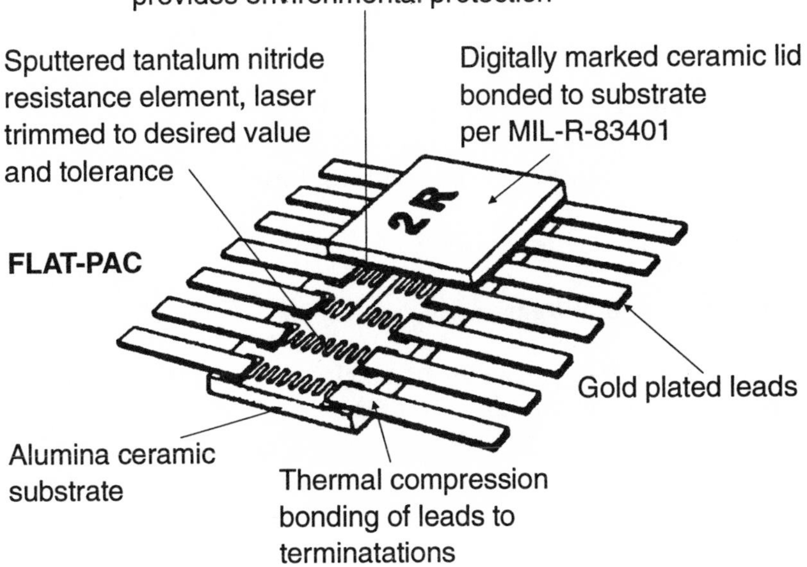

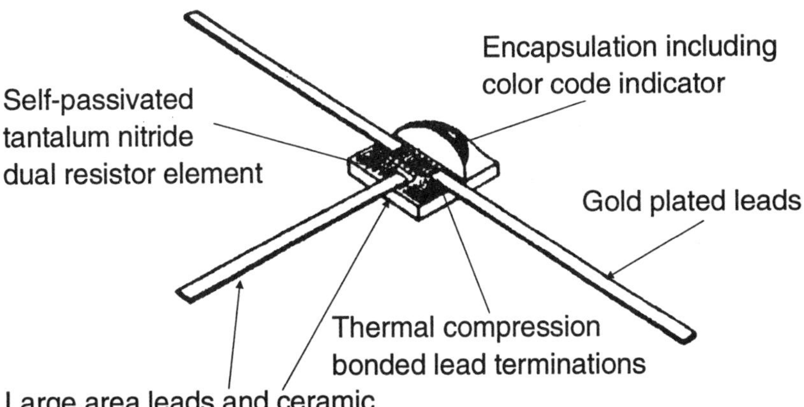

Resistor Network Package Options
Figure 9.25b

ADVANTAGES OF RESISTOR NETWORKS COMPARED TO DISCRETE RESISTORS

- Fewer components are handled
- Assembly time is faster
- On-board space and over-all package weight is reduced
- Performance parameters are improved
- Resistor characteristics have greater uniformity
- Circuit stability is improved
- Superior temperature tracking is achieved

The above advantages provide greater circuit reliability and reduced assembly and inventory costs. (See Figure 9.26)

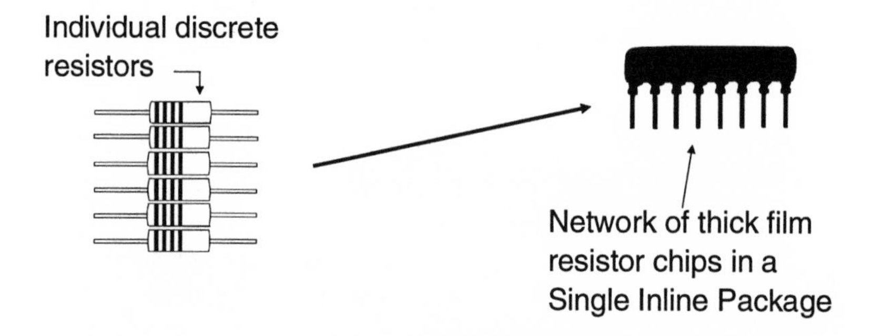

Resistor Network Replacing Individual Discrete Resistors
Figure 9.26

TYPICAL RESISTOR NETWORK APPLICATIONS

- Standard and custom precision voltage divider networks
- Analog to Digital (A/D) convertors and
  Digital to Analog (D/A) convertors
- Resistor arrays for logic gate circuitry
- Resistor arrays for balancing and terminating transmission lines
- Custom resistor circuit interconnect patterns

ATTENUATOR CIRCUITS

An **attenuator**, or voltage reduction circuit, is a resistor network that reduces the amplitude of a voltage or electrical signal without introducing any appreciable distortion or change in the shape of that signal. In communication systems, attenuators are used to reduce the strength of a signal.

An attenuator network is available as either a *fixed attenuator*, a *step attenuator*, or a *variable attenuator*. Examples of fixed attenuators are shown in Figures 9.27 and 9.28.

FIXED ATTENUATORS

A *T-Pad* or *T-Section* attenuator (Figure 9.27) provides a desired value of voltage reduction for a signal while maintaining consistent input and output resistance. It may also be designed to match unequal source and load resistances. Any value of attenuation may be selected by varying the values of the resistors.

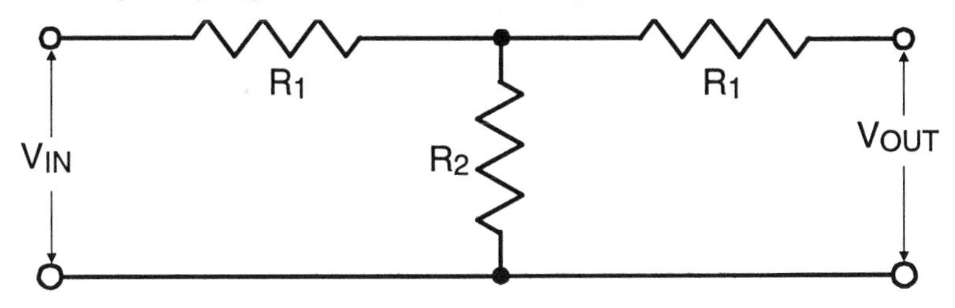

T-Pad or T-Section Attenuator
Figure 9.27

The *L-Type* Single Section attenuator (Figure 9.28) is used when the output resistance and the input resistance of the attenuator differ in their value.

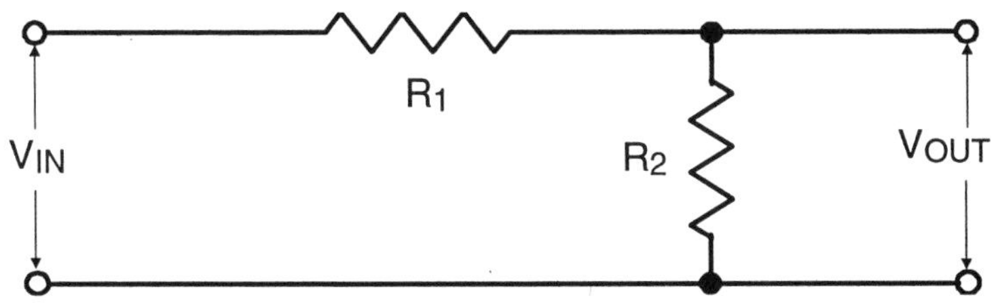

L-Type Single Section Attenuator
Figure 9.28

STEP ATTENUATOR

A group of two or more fixed resistors connected in series is referred to as a *resistor chain* or *voltage divider*. The *step attenuator* section of the circuit shown in Figure 9.29 can be packaged either as a group of individual discrete resistors connected in series or as a resistor network in a single package connected in series.

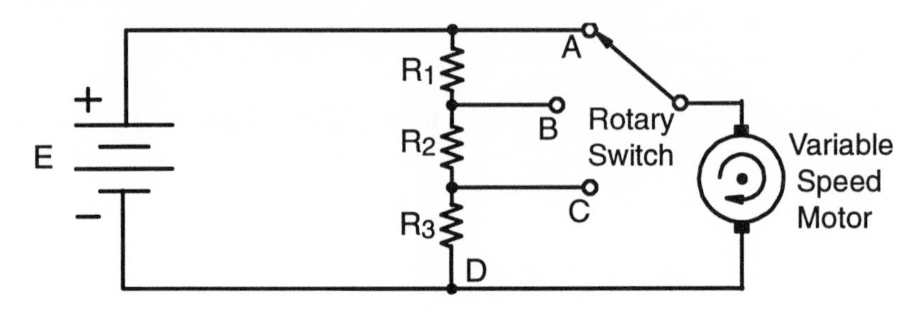

Step Attenuator - With Three Fixed Resistors
Figure 9.29

The resistors act to divide the voltage source, E, into specific values of voltages. The selected setting of the single pole, triple position rotary switch acts to apply the desired voltage to the load (the variable speed motor). (See Figure 9.29)

- With the switch's moving contact at point A of the resistor chain, the total supply voltage, E, is connected to the load.
- When the moving contact of the switch is moved to point B , the voltage at the load is reduced by the voltage across resistor R_1.
- As the switch is moved from one point to another, the selected voltage applied to the motor depends upon the voltage in the chain between the moving contact terminal of the switch and point D, the negative terminal of the voltage source.

In this circuit, the resistors are functioning as a voltage divider network being used as a step attenuator. The voltage selection (speed selection) for the motor depends on the supply voltage, E, and the setting of the switch. The available motor speeds are preset by the value of each resistor in the chain.

The number of speed selections is determined by the number of positions on the rotary switch (the same amount as the amount of resistors in the voltage divider).

VARIABLE ATTENUATOR

In the circuit of Figure 9.30, the potentiometer connected across the supply voltage, E, is functioning as a variable attenuator to adjust the speed of a variable speed DC motor.

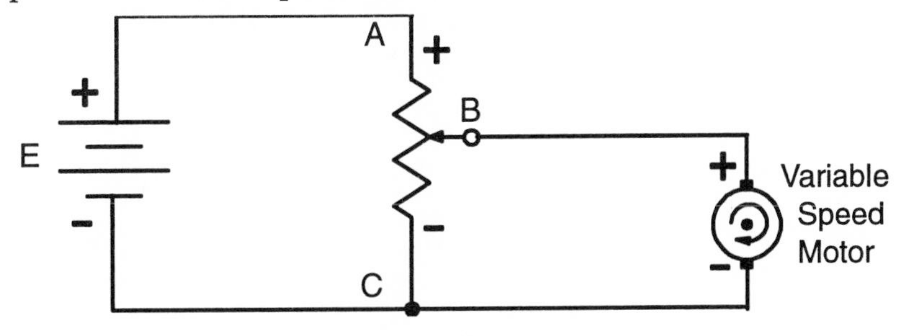

Variable Attenuator (Potentiometer)
Figure 9.30

Any part of the supply voltage can be applied across a variable speed motor by connecting the potentiometer's moving control terminal, B, to the positive terminal of the load. The voltage applied to the motor depends upon the setting of the potentiometer and determines the speed of the DC motor.

The supply voltage, E, connected across the outer terminals of the potentiometer (points A and C), is attenuated by the appropriate setting of the potentiometer. The user may select any motor speed and is not limited to a predetermined number of selections.

The same principle of using a potentiometer as a variable attenuator is shown in a volume control application in Figure 9-31.

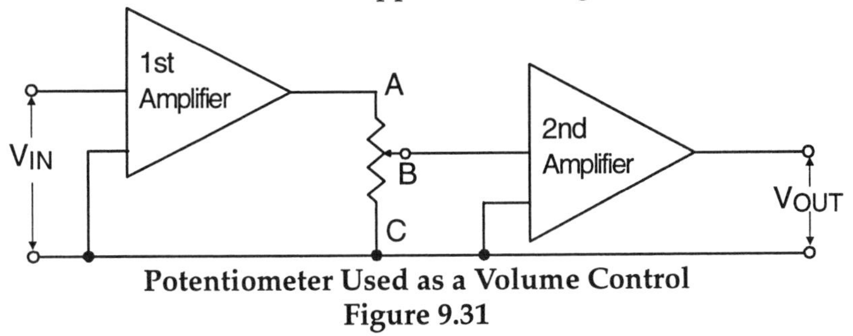

Potentiometer Used as a Volume Control
Figure 9.31

The output of the first amplifier is connected across the potentiometer terminals A and C. The moving contact terminal, B, is connected to the input of the second amplifier. The setting of the potentiometer will determine the volume at the output of the system.

REINFORCEMENT EXERCISE

Answer TRUE or FALSE

1. Resistors provide: current adjustment, voltage division, attenuation, part or all of a circuit load, and part of the elements of a timing network.

2. The important parameters to specify fixed resistors are: ohmic value, tolerance, power capability, temperature coefficient, operating and storage temperature range, resistance to humidity, and the ability to withstand other environmental stresses.

3. Carbon composition resistors can be used where instantaneous surge currents are present since they are capable of handling sudden overloads of current and power for a short time.

4. Metal film resistors are specified for circuit applications requiring tight tolerances and relatively low temperature coefficients.

5. For power applications, nichrome wirewound resistors will generally offer the greatest power capability.

6. For precision uses, wirewound resistors with extremely tight tolerances and very low temperature coefficients are available.

7. A thermistor is a special resistor with a zero TC characteristic and will not change in ohmic value as temperature changes.

8. Thick film resistor chips provide superior temperature tracking and lower TC characteristics than do thin film resistor chips.

9. A single-turn potentiometer or trimmer provides better resolution and precision than a multi-turn potentiometer or trimmer.

10. A trimmer is generally designed to be adjusted with a screwdriver and does not normally accept a knob as an adjustment device.

11. A rheostat is a variable resistor with two terminals. A potentiometer can act as a rheostat by using only two of its three terminals (the control terminal and either end terminal).

12. An attenuator circuit is used to reduce the level of a voltage source to a specified lower level at its output terminals.

Determine the value and tolerance of each resistor:

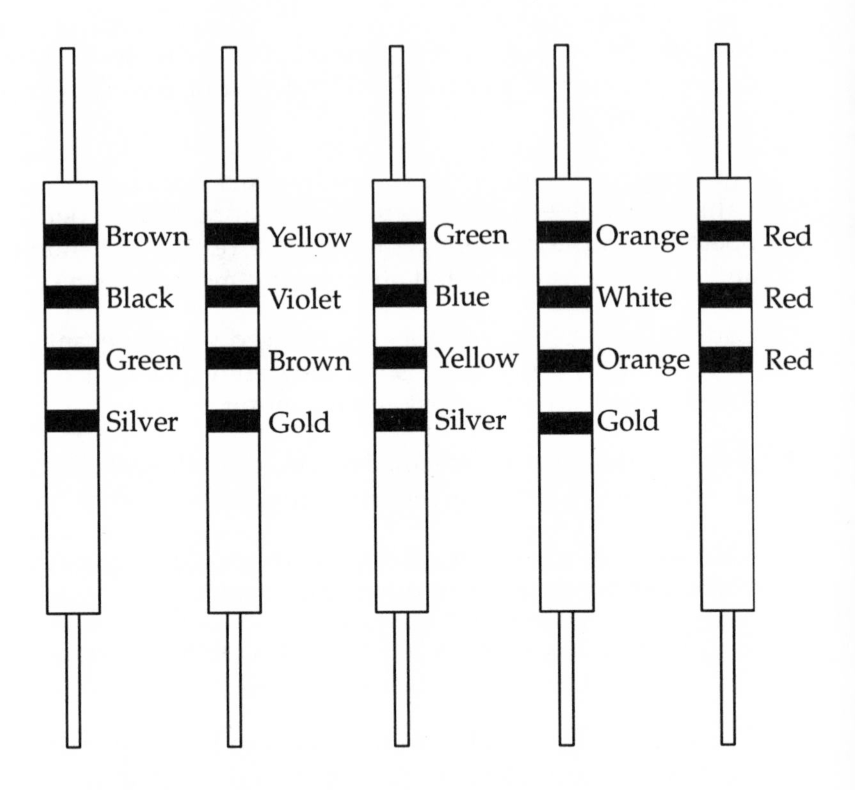

The answers to this reinforcement exercise are on pages 275-276.

10 CHAPTER TEN

CAPACITORS

CAPACITORS

A **capacitor** is an electrical storage component that has the capability of accepting an electrical charge (voltage and current) from a voltage source. This charge can be stored for as long as required and then released. The unit of capacitance is the *farad*, F, named after Michael Faraday. By definition, a one farad capacitor will charge to one volt in one second with a charging current of one ampere.

In its basic form, a capacitor consists of two conducting metal plates with terminals attached to each plate. Sandwiched between the two metal plates is a nonconducting material called the *dielectric*.

Capacitance is the measure of the storage capability of a capacitor, and its value is dependent upon the area of the metal plates, the distance between them, and the specific dielectric material used.

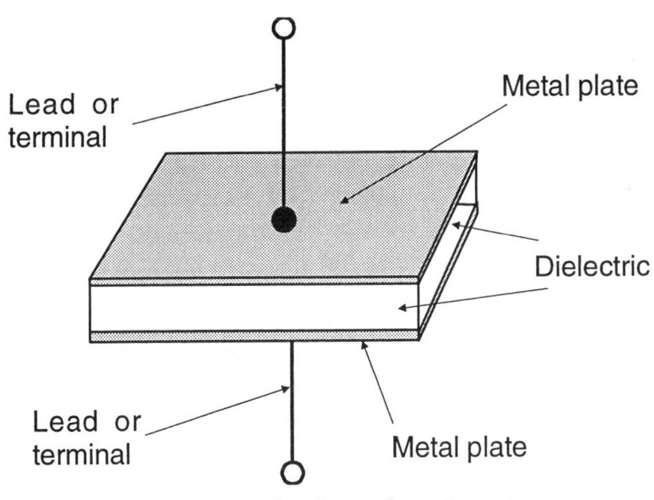

Concept of a Capacitor Structure
Figure 10.1

The insulating materials generally used as dielectrics include:

- Aluminum oxide
- Paper plus Mylar™
- Paper plus polypropylene
- Polysulphone
- Tantalum oxide
- Ceramic
- Silicon dioxide (glass)
- Polycarbonate
- Mica
- Mylar™
- Teflon™
- Air

Typical values of capacitance are fractions of a farad and are listed below:

- One millionth of a farad = 1 microfarad (1 μF) = 1×10^{-6} farads
- One trillionth of a farad = 1 picofarad (1 pF) = 1×10^{-12} farads

A 1.0 farad capacitor is considered to be a huge value of capacitance and is a rarely used circuit component. It is manufactured, however, for special circuit applications where extremely high capacitance values are required.

CAPACITOR APPLICATIONS DEFINITIONS

DC FILTERING - Changes pulsating DC voltage to steady-state DC voltage.

AC COUPLING and DC BLOCKING - Transfers an AC signal from the output of an amplifier to the input of a succeeding amplifier stage. At the same time, a capacitor blocks the steady-state DC voltage at the output of the amplifier stage from getting to the input of the next amplifier stage.

VOLTAGE SENSING - Acts to acquire a voltage, store it for as long as desired, and then make the voltage available to perform a specific function. A typical use is in a voltage measurement application or sample and hold circuit.

PHASE SHIFTING - Acts to change the time relationship among voltages of the same frequency. Phase shifting networks consist of capacitors, inductors, and/or resistors in passive delay lines for both linear and digital communications systems.

TIMING - A capacitor in conjunction with a resistor forms a resistor/capacitor (RC) network to provide the required length of time to charge or discharge a voltage across a capacitor.

TUNING - Acts to select a specific frequency among multiple frequencies in conjunction with an inductor or resistor.

HIGH FREQUENCY BY-PASS - Provides a low resistance path for high frequency signals. Generally used in conjunction with an RF choke to keep RF voltages out of a common DC power supply or other low frequency circuits.

CAPACITOR TYPES

Capacitors are categorized as *variable, non-polarized,* and *polarized.* Their schematic symbols are shown in Figures 10.2. Within each category are many varieties of capacitance, tolerance, voltage capability, TC, temperature range, the capability to withstand environmental stress, package configurations, and process techniques. Different types of dielectric materials are used to obtain the unique electrical characteristics required for each type.

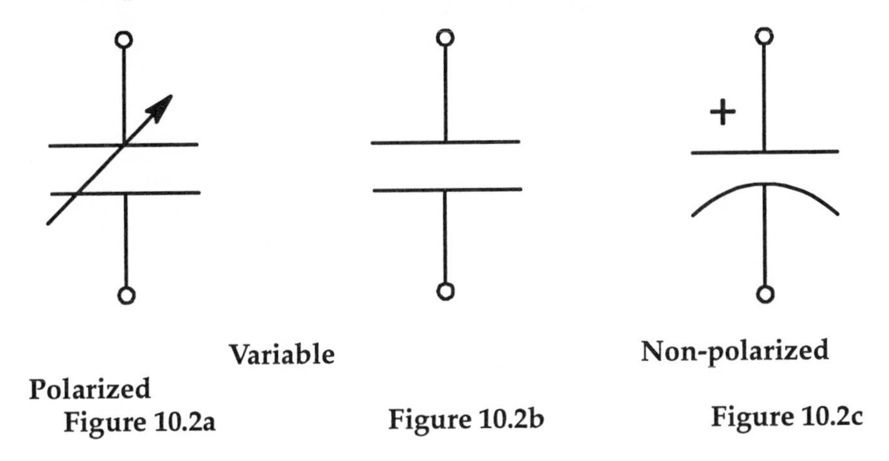

Variable **Non-polarized**

Polarized

Figure 10.2a **Figure 10.2b** **Figure 10.2c**

VARIABLE CAPACITORS

Variable capacitors have the capability of value adjustment. Air is usually used as the dielectric.

An air dielectric variable capacitor consists of two sets of equally-spaced aluminum plates electrically insulated from each other. One set of plates, fixed in a frame, intermeshes with a second set of plates mounted on a rotatable shaft with air between the plates. When the shaft is rotated, one set of plates meshes between the other set, changing the area of the meshed plates, producing a change in capacitance. Mica or mylar separators are often placed betwen the aluminum plates to prevent them from touching each other.

When used with an inductor, a variable capacitor acts as the tuning component of a *tuned* or *resonant circuit* used to select, or tune to, a specific radio frequency (radio station or TV channel). Air dielectric variable capacitors range in value from about 2 pF up to several thousand pF with voltage ratings to several thousand volts. Variable capacitors are categorized as non-polarized types.

NON-POLARIZED (NP) CAPACITORS

The NP capacitor will accept either plus or minus DC voltages or AC voltages across its terminals and is manufactured in a variety of dielectric materials. Different dielectric materials account for the specific characteristics of the non-polarized capacitor and identify the many different types that are manufactured, as follows:

METALIZED POLYESTER (Mylar™) - used for mid-frequency coupling and filter applications, timing circuits, and mid-frequency tuning circuits. Capacitance values range from 1,000 pF to 18 μF. Breakdown voltage ranges from 50 to 400 volts.

PAPER PLUS MYLAR OR PAPER PLUS POLYPROPYLENE - used for the same applications as Mylar. Capacitance values range from .001 μF to 500 μF. Breakdown voltage ranges from 100 to 500 volts.

CERAMIC, MICA, AND GLASS - used for high-frequency coupling and filter circuits, fast timing circuits, and RF tuned circuits. Capacitance values range from 1pf to 10 μF with voltage breakdown capability up to 50,000 volts.

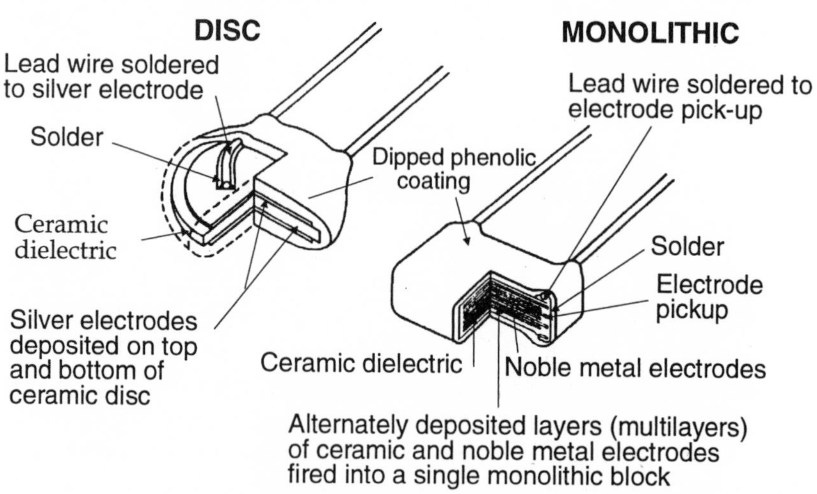

Typical Construction Features of Ceramic Capacitors
Figure 10.3

TEFLON™, POLYCARBONATE, POLYPROPYLENE and POLYSULPHONE (FOIL) - used for high temperature, high frequency, and high reliability applications. Capacitance values range from 10 pF to 100 μF with voltages ranging from 50 to 400 volts.

POLARIZED CAPACITORS

A *polarized* capacitor must have the proper voltage polarity applied across its terminals; if incorrect polarity is applied, the device could be destroyed. Because of this characteristic, an AC voltage cannot be applied across a polarized capacitor unless a DC offset voltage also exists in series with the AC. The amplitude of the DC offset voltage must be greater than the AC voltage with the correct voltage polarity always maintained across the capacitor terminals. Either aluminum oxide or tantalum oxide is used as its dielectric.

A polarized capacitor has about 100 times the capacitance of a non-polarized type with the same voltage rating in a similar-sized package. For applications where a large value of capacitance is needed in a small package, a polarized capacitor will generally be chosen if the circuit design is compatible with this selection.

ALUMINUM OXIDE (ALUMINUM ELECTROLYTIC) - used over a limited temperature range (-40°C to +85°C for commercial, industrial, and consumer applications) in DC filters, low frequency AC filters, and voltage storing circuits. Capacitance values range from 0.1 μF to 1 farad. Voltage ratings range from 3 to 500 volts.

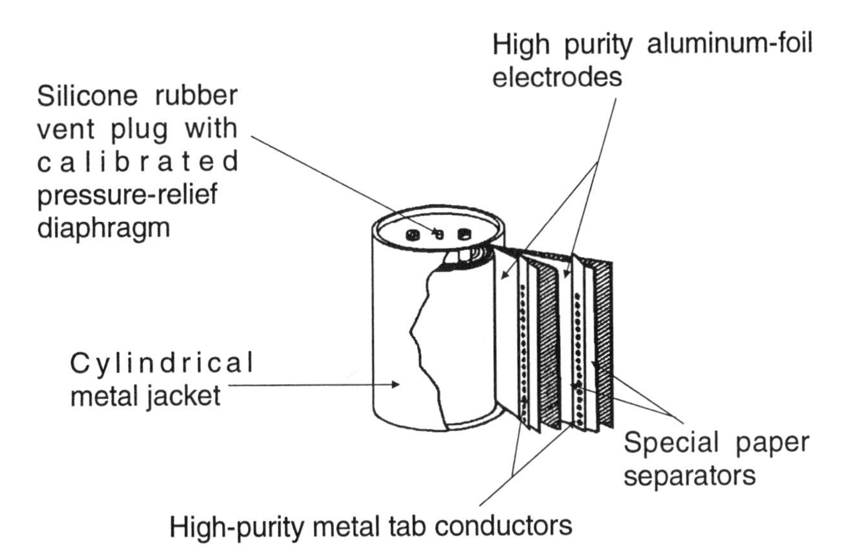

Construction Features of Aluminum Electrolytic Capacitors
Figure 10.4

TANTALUM OXIDE - (FOIL, SOLID, AND WET SLUG TYPES) - used for the same applications as aluminum oxide types having approximately the same capacitance range and voltage ratings. Tantalum oxide capacitors are specifically intended for military and space applications because of their wide operating and storage temperature range (-55°C to +125°C). In comparison with aluminum oxide capacitor types, they have greater stability and reliability, are physically smaller, but are more expensive.

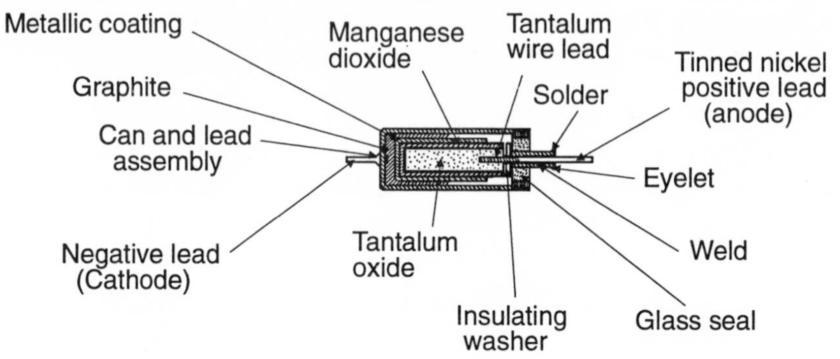

Construction Features of a Typical Solid Tantalum Capacitor
Figure 10.5

Both aluminum oxide and tantalum oxide capacitors are normally polarized types. They are also manufactured in non-polarized versions for special applications where the requirement for non-polarity applications is critical and large values of capacitance are required. Non-polarized aluminum and tantalum capacitors are made in somewhat larger package sizes than the equivalent capacitance/voltage combination of polarized types.

CHARGING AND DISCHARGING

The charging and discharging action of a capacitor can be compared to the filling and emptying of a water bucket. A bucket can be filled with water, and assuming there is no leak in the bucket, the water will be stored there for as long as desired. When needed, the bucket can be emptied and the water used to perform some function. The bucket can be filled instantantly by placing it under a waterfall or be filled very slowly by putting it under a slow-running water tap. It can be emptied instantaneously by turning the bucket over or be emptied very slowly by tilting it slightly, allowing the water to trickle out. The rate of filling and emptying is, therefore, determined by the specific technique that is chosen.

CHARGING A CAPACITOR

Conceptually, charging a capacitor works the same way as filling a water bucket. The time it takes to fill the water bucket will depend on the manner in which it is filled. A charging voltage and current can be applied to a capacitor to "fill it" electrically.

When fully charged (or filled), the voltage (V_C) across the capacitor will equal the magnitude of the charging voltage and will have the same polarity. The time it takes to charge will depend on the value of the capacitor and resistor (RC network).

In the circuit diagram (Figure 10.6), the capacitor (C) is being charged by the voltage source (E) with the capacitor in series with resistor (R_S). The resistor, in ohms, and the capacitor, in farads, combine to form a resistor-capacitor network whose product, $R_S \times C$, equals one time-constant (in seconds). The required time-constant is determined by the circuit application.

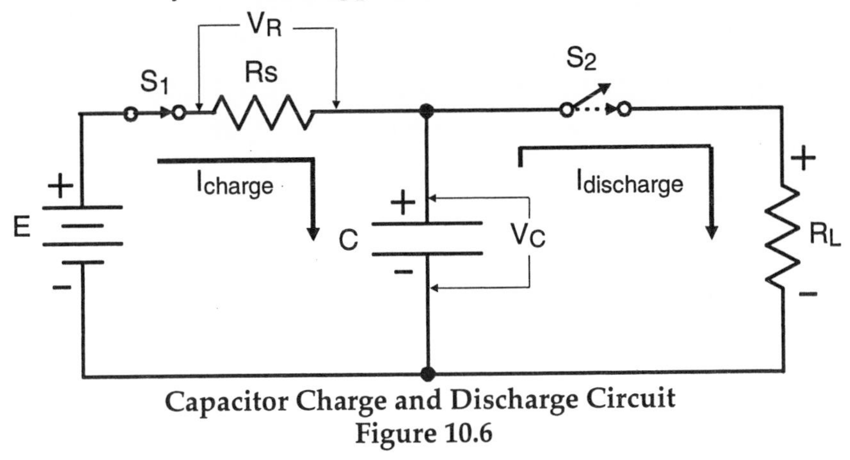

Capacitor Charge and Discharge Circuit
Figure 10.6

CHARGING SEQUENCE:

- When switch S_1 is closed and switch S_2 is open, the capacitor will start charging to 63% of the supply voltage during the first time-constant.

- After the second time-constant, the capacitor will be additionally charged to 63% of the remaining supply voltage.

- After the third time-constant, the capacitor will be further charged to 63% of that remaining voltage, and so on in this manner, until the capacitor becomes fully charged.

Six RC time-constants are necessary to fully charge the capacitor to the magnitude of the supply voltage. If the supply voltage is 100 volts, and one time-constant is 1 second, after the first time-constant (1 second) the capacitor will be charged to 63 volts.

After 2 seconds the capacitor will be charged to 86.3 volts (63 volts plus 63% of 37 volts), etc. After 6 seconds, the capacitor voltage will be 99.75 volts (essentially fully charged). (See Figure 10.7)

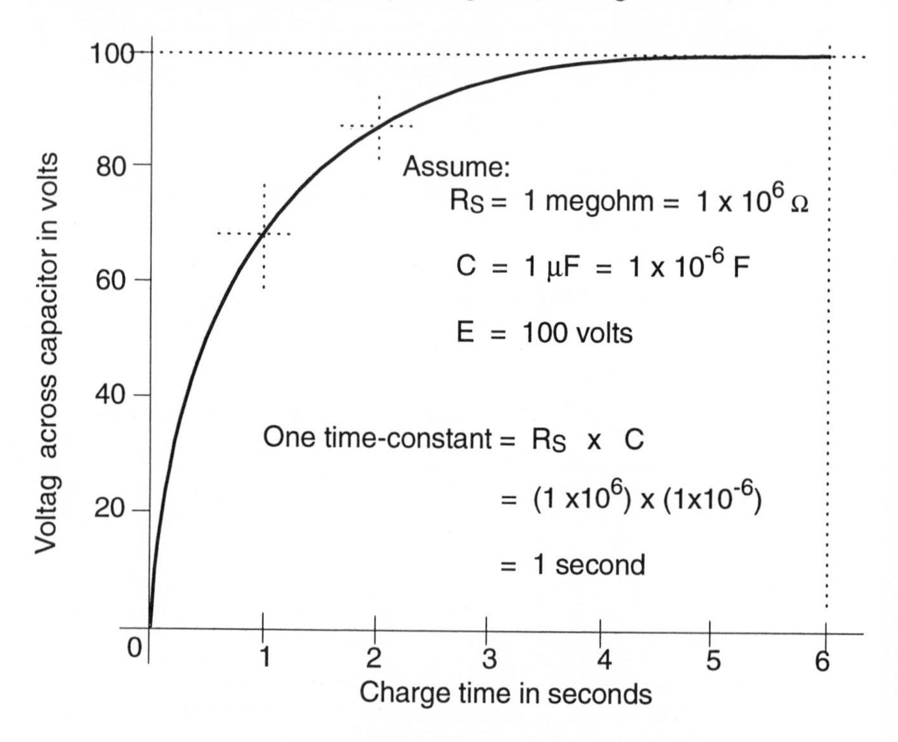

Voltage Across Capacitor vs. Charge Time
Figure 10.7

- The nature of the charging waveshape, called an *exponential curve*, depends on the manner in which a capacitor charges and is true for all capacitors. Because of its exponentially charging characteristic, a capacitor will never fully charge to the exact value of the applied voltage.

- At the start of the charging sequence, the entire supply voltage is across the series resistor, and no voltage exists across the capacitor. The charging current (I_{charge}) is equal to the voltage (V_R) across R_S divided by R_S. ($I_{charge} = V_R/R_S$)

- Since this is a series circuit and the source voltage remains constant, as the voltage across the capacitor gradually increases, the voltage across resistor, R_S, decreases. With the charging current equal to the voltage across resistor, R_S, divided by its resistance, the charging current will decrease and the rate at which the capacitor is charging will decrease. The voltage across the capacitor will increase at a gradually decreasing rate.

- Even though the rate of charge is continuously decreasing, after six time-constants (6 x RC), the capacitor is essentially fully charged, and the charging current ceases. With zero current flowing in the capacitor, its resistance is infinite.

A second circuit connected in parallel with the capacitor will have the same voltage (V_C), since all voltages in parallel are equal.

To increase the charge time, either the value of the series resistor or the value of the capacitor can be increased, or both values can be increased. To decrease charge time, the value of either the series resistor or the capacitor can be decreased, or both values can be decreased. In most cases, it is easier to adjust the value of the resistor.

The value of the resistor can also be decreased to zero (no resistor in the circuit). In this case, since R is equal to zero, the R x C value is zero and 6 x R x C is still zero. Under this condition, the capacitor will charge instantly to the magnitude of the applied voltage.

The time for charging the capacitor can range from zero time (R=0) to as long as needed, depending on the practical values of R and C.

In principle, a capacitor and a rechargeable battery function somewhat in the same manner. When initially applying a charging voltage to a discharged battery, charging current is high.

As the charge in battery increases, charging current decreases, reducing the rate of charge until the battery becomes fully charged. When this occurs, the charging current is essentially zero and charging of the battery stops. With a lead acid battery, a regulator is required to prevent overcharging that could result in damage to the battery.

A battery, however, will eventually lose its capability to be recharged. A capacitor, on the other hand, can be discharged and recharged indefinitely as long as its ratings are not exceeded.

DISCHARGING THE CAPACITOR

Assuming that there is no electrical leakage in the capacitor, the energy, or charge, in the capacitor will be stored for as long as desired. Realistically, the charge will be maintained for a finite interval of time depending on the particular capacitor's dielectric material.

When needed, the capacitor can be "emptied", or discharged, by closing switch S_2 in the circuit of Figure 10.6, and opening switch S_1, thereby connecting the load (R_L) across the capacitor. The discharging and decreasing current through the load will produce a decreasing voltage ($I_{discharge} \times R_L$) across the load. To fully discharge the capacitor, the time ($6 \times R_L \times C$) will depend on the value of the capacitor and the value of R_L in the discharging path.

Note that the voltage across the charged capacitor is discharged in a manner that is opposite to the charging manner. (See Figures 10.8)

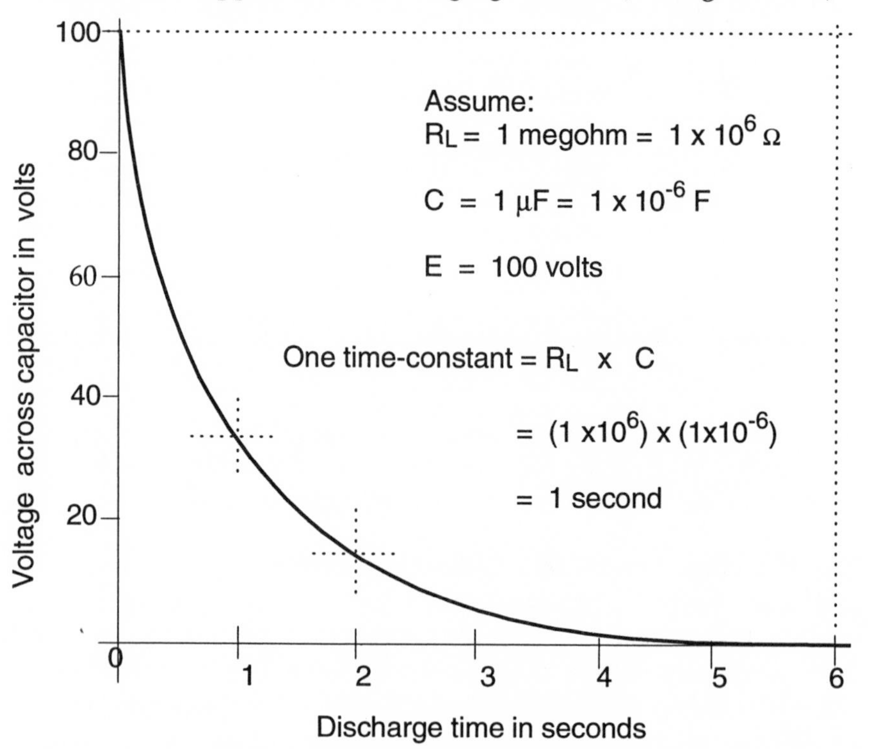

Voltage Across Capacitor and Load vs. Discharge Time
Figure 10.8

DISCHARGING SEQUENCE:

With switch S_1 open and switch S_2 closed, the capacitor will discharge in a timing sequence similar to the charging sequence.

- After the first time constant, 63% of the voltage in the charged capacitor will be discharged.
- After the second time constant, 63% of the remaining voltage will be discharged.
- This action will continue in this manner until six time constants have elapsed, essentially fully discharging the capacitor. The capacitor will never fully discharge; an extremely small voltage, called *dielectric absorption*, will still remain across the capacitor.

Depending upon the values of R_L and C, the discharge time can be as long or as short as required.

Normally, when maintenance personnel are troubleshooting a circuit containing capacitors, as a safety precaution, the capacitors should be discharged by using a shorting bar (a screwdriver or short wire of zero resistance) before attempting to work on the circuit. (It is often safer to use a low value resistor as a discharging device.) Neglecting to discharge the capacitor(s) might result in damage to equipment or harm to personnel.

The circuit shown in Figure 10.6 requires manual actuation of the two switches S_1 and S_2 and is not a practical circuit. This technique is used, however, to illustrate the charging and discharging action of a capacitor. More appropriate techniques of charging and discharging action are shown later in this chapter in the DC filtering and photoflash applications.

RESISTOR-CAPACITOR (RC) NETWORKS

In Figure 10.6, the charge and discharge time constants are shown as being equal since the series resistor (R_S) and the load resistor (R_L) are equal to each other and a common capacitor (C) is used for both charge and discharge circuits. In an actual application, these components can have any value to achieve a desired time-relationship between the charge and discharge of a capacitor.

A resistor-capacitor (RC) network provides the timing components of an electronic timing circuit. The charge and discharge action of a capacitor establishes the method and principle on which many capacitor applications are based. As the value of either the resistor or capacitor (or both) of the RC network is changed, the charge and discharge times are changed accordingly.

RC networks are easy to adjust and, with additional circuitry, are often less expensive than the more complex mechanical timing mechanisms. They do not wear out and generally sustain a level of circuit stability and reliability that is not readily duplicated by some equivalent mechanical timing mechanisms.

TYPICAL CAPACITOR APPLICATIONS

DC FILTERING

DC filtering is the process of changing pulsating DC voltage to steady-state DC voltage, the final source of voltage for electronic equipment. Before DC filtering can occur, an AC voltage source used as the prime source of power must first be changed to pulsating DC voltage. This intermediate process is referred to as *rectification* and is explained in detail in Volume Two - Part One - Discrete Semiconductors.

After either half-wave or full-wave rectification, the resulting pulsating DC voltage is changed into steady-state DC voltage. Because of its charge, storage, and discharge capability, a capacitor in conjunction with resistance is used to provide DC filtering.

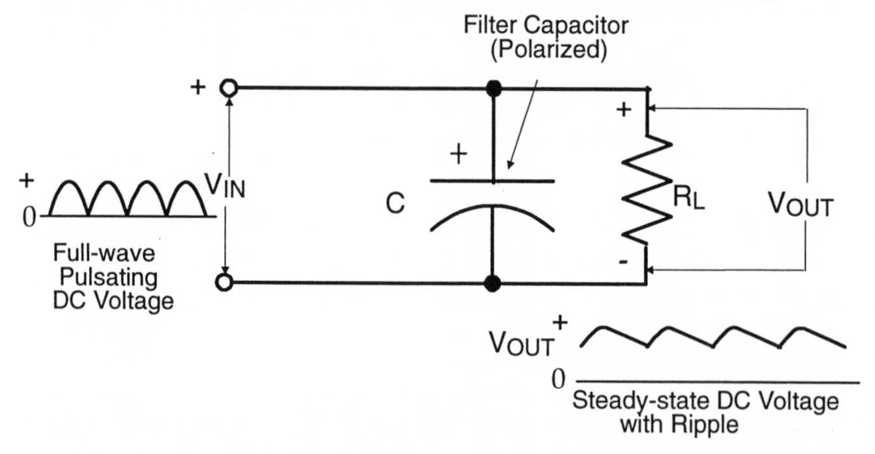

DC Filtering - Changing Pulsating DC to Steady-state DC
Figure 10.9

In this application, full-wave pulsating DC voltage is applied to the input of the DC filter. During the charging interval, with no series resistor, $R_S \times C_1$ = zero and the capacitor charge time is zero.

- The capacitor charge follows the increasing input voltage (V_{IN}) with no delay, charging the capacitor to its peak applied voltage. As the input voltage decreases to zero, the voltage across the capacitor discharges slowly through the load, with the discharge time determined by the time-constant ($R_L \times C$).

- The next charging pulse will recharge the capacitor to the peak applied voltage before the capacitor voltage has fully discharged. Each succeeding charging pulse will charge the capacitor to the peak applied voltage then allow the capacitor to discharge slowly until the next charging pulse comes along. (See Figure 10.9)

- The DC output voltage (V_{OUT}) is called steady-state DC since it never decreases to zero but has a finite average value. The DC output voltage level, however, is not a constant value but varies with each succeeding input charging pulse and decreasing discharging voltage. This voltage variation is called *ripple*, an undesirable characteristic and should be eliminated or minimized. (See Figure 10.9)

- One technique of minimizing ripple, or effectively eliminating it to produce a constant output voltage (steady-state DC without ripple), is to add one or more filter sections (series resistor R_1 and filter capacitor C_2) to the circuit to improve the over-all DC filtering action. (See Figure 10.10)

Note that for DC filters, high capacity polarized capacitors are used to create a large RC time-constant during the discharge portion of the filtering process.

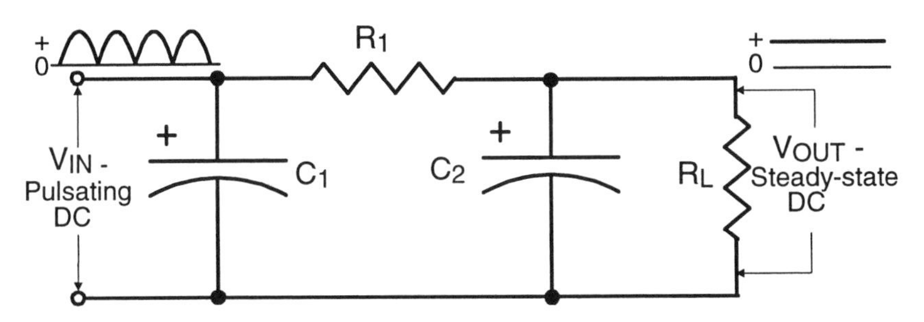

A Multi-section DC Filter to Minimize or Eliminate Ripple
Figure 10.10

AC COUPLING AND DC BLOCKING

A capacitor provides a frequency-dependent function. Its AC resistance is dependent on several factors: its capacitance in farads, the frequency of the applied AC voltage in hertz, and a mathematical constant that is equal to 2 x π (2 x 3.14 = 6.28).

This AC resistance is referred to as **capacitive reactance** and is frequency dependent in value. The letter **X** designates the reactance, measured in ohms. Capacitive reactance, symbolized as **X_C**, is stated mathematically as:

$$X_C = \frac{1}{2\pi fC} \quad \text{in ohms}$$

where: π is the constant = 3.14
f is the frequency of the of the AC voltage in hertz
C is the value of the capacitor in farads

When used as an interstage coupling capacitor (see Figure 10.11), the capacitive reactance (X_C), must be made as small as possible, compared to the load resistance (R_L). A high value of C will minimize any attenuation or reduction of the signal being coupled (transferred) from the output of the first amplifier (shown in block form) to the input of the next one. The load resistance (R_L), represents the input resistance of the succeeding amplifier.

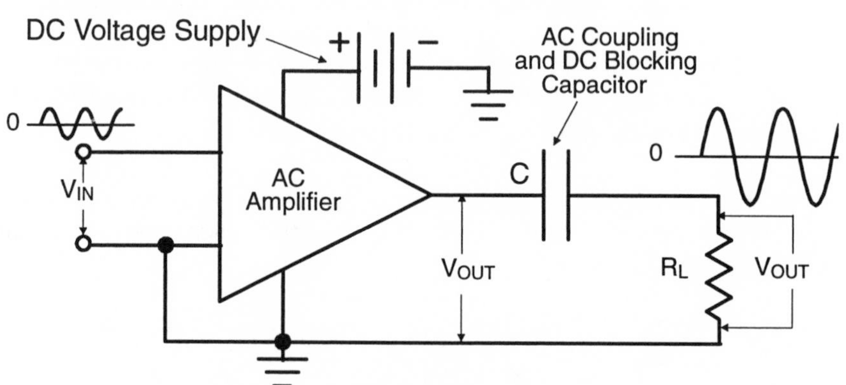

AC Coupling and DC Blocking
Figure 10.11

The constant, 2π (6.28), and the frequency (f) of the amplified signal are elements of the formula for capacitive reactance and cannot be modified. The value of the capacitor (C) can be selected to ensure that it is big enough to make its reactance (X_C) extremely small compared to the resistance of the load (R_L). A very small X_C will act to couple the amplified AC signal at the output terminal of the amplifier to the load (R_L)) without signal attenuation (reduction).

Several different voltages can exist at the output terminal of an amplifier. In Figure 10.11, two voltages are present at the output terminal, one being the amplified AC signal that has been coupled to R_L, and the other, a portion of the steady-state DC supply voltage connected to the power input terminal of the first amplifier through the amplifier's internal circuitry.

The DC supply voltage is necessary to enable the amplifier to provide amplification of the AC signal. This DC voltage, or any portion of it, must be blocked from reaching the load, R_L, to avoid disturbing the operating conditions of the second amplifier. While efficiently coupling the amplified signal to R_L, the capacitor will also act to block DC from R_L. Since the dielectric of a charged capacitor has essentially infinite DC resistance, the capacitor acts as an open circuit to DC, allowing no DC voltage to be applied across R_L.

Anther way of viewing this DC blocking effect of the capacitor is to note that the frequency of steady-state DC voltage is zero. If the value of the frequency (f) is zero in the equation for X_C, then X_C is equal to infinity, effectively blocking the DC from getting to the load.

RADIO FREQUENCY (RF) BYPASS APPLICATION

In the circuit of Figure 10.12, a capacitor provides a low resistance path for any high frequency AC that is present at the DC power supply terminal of the RF amplifier (shown in block form).

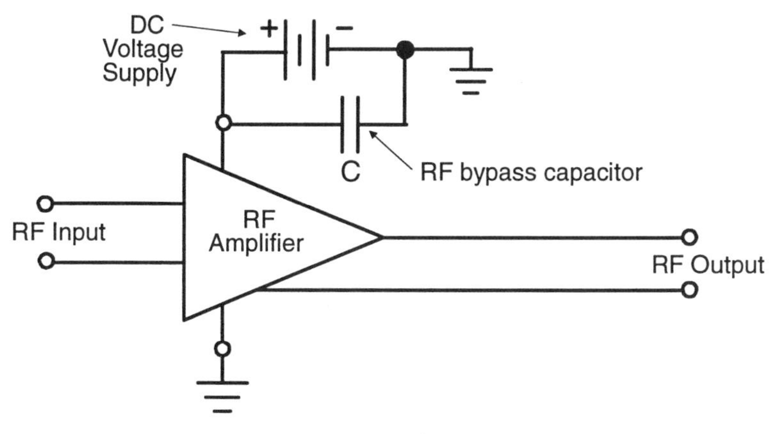

RF Bypass Application
Figure 10.12

RF voltages that exist in one amplifier need to be prevented from interacting with a common power supply of an entire system. For high frequency (RF) amplifiers, the capacitive reactance (X_C) of a bypass capacitor is very low compared to the resistance of the supply voltage and any interconnecting circuits. Any RF voltage present at the voltage supply terminal of an amplifier will be *bypassed* (shunted) by the capacitor around the supply and circuit conductors.

To bypass high frequency RF to ground, a small capacitor is incorporated in many printed circuit board applications. The capacitor bypasses unwanted high frequency signals around critical sections of a circuit and assures circuit stability by maintaining a high level of separation between between these different sections.

PHOTOFLASH (ENERGY STORAGE) APPLICATION

In the photoflash circuit of Figure 10.13, the capacitor, C, is used to store a charge when the voltage (E) is applied.

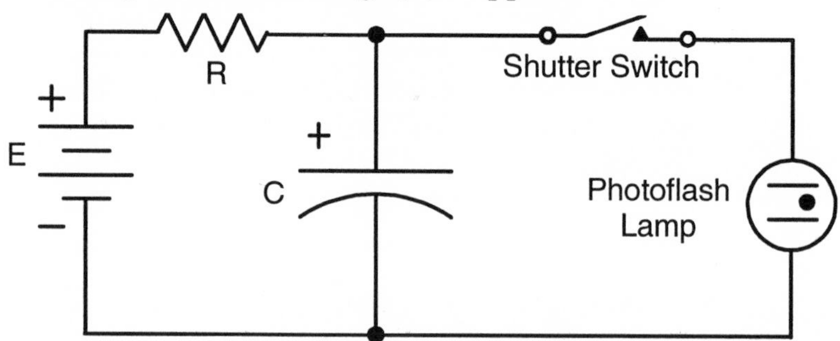

Photoflash Application
Figure 10.13

When the shutter switch is depressed, the photoflash lamp circuit is closed, placing the charged capacitor in parallel with the photoflash lamp. If the voltage across the capacitor is high enough, it will ionize (break down) the gas in the photoflash lamp, changing its initial infinite resistance to zero, to provide a discharge path for the capacitor. The current through the photoflash lamp will provide the required light intensity for proper light exposure.

This circuit is a simplification of the actual circuit used in photoflash equipment, but the principle involved is the same. It illustrates the charge, storage, and discharge of the capacitor used in this application. Note that a polarized capacitor is used to provide the large capacitance required in this circuit.

CAPACITOR SPECIFICATIONS

A capacitor is specified by its electrical, thermal, and physical characteristics as well as by its ability to withstand the environmental stresses to which it may be subjected.

NOMINAL CAPACITANCE VALUE (C) - Specified at +25°C

CAPACITANCE VALUE TOLERANCE - This specification is expressed as a plus or minus percentage of the nominal value of the capacitor at +25°C.

MAXIMUM VOLTAGE RATING - This is the maximum DC voltage that can be applied across a capacitor for continuous operation, specified at both +25°C and at the maximum rated temperature. If this voltage is exceeded for even a short time, it may arc through the dielectric and destroy the capacitor. Unlike a resistor that dissipates power (voltage times current), the capacitor has a dielectric as part of its structure. The dielectric, a nonconductive material, has essentially infinite resistance when the capacitor is charged and no DC will flow into or through the capacitor. Power capability of a capacitor is not specified, since little or no power is being dissipated.

TEMPERATURE COEFFICIENT (TC) - This is expressed either as parts per million per degree C (ppm/°C), or as the percent change in capacitance per °C, for a specified temperature range. In the capacitor equivalent circuit of Figure 10.14, the variable capacitance, shown in parallel with capacitor C, represents its temperature coefficient. A change in temperature causes a change in the total capacitance value according to its temperature coefficient.

OPERATING AND STORAGE TEMPERATURE RANGE -

- For military and space equipment:
  -55°C to +125°C

- For commercial, industrial, and consumer use:
  Any range specified by the manufacturer, typically, 0°C to 85°C

ENVIRONMENTAL REQUIREMENTS - The selection of a capacitor package is dictated by the ability of the capacitor to withstand the stresses of vibration, humidity, salt spray, fungus, chemical solvents, and other environmental conditions. For the stresses inherent in military and space applications, a hermetically sealed package must be used to meet the applicable military specifications.

EQUIVALENT SERIES RESISTANCE (ESR) - This is a function of the resistance of the plates, the lead material, and the construction of the capacitor. The lower the plate resistance and lead resistance of the capacitor, the lower the value of the equivalent series resistance. It is shown in the equivalent circuit of Figure 10.14 as a small resistor in series with the capacitor and represents an energy loss (I^2R) or a voltage loss (I x R) caused by the charging current.

LEAKAGE RESISTANCE or LEAKAGE CURRENT - This indicates the quality of the dielectric. The fewer contaminants in a dielectric, the higher the leakage resistance or the lower the leakage current. Leakage current can be compared to the very small amount of water that may be leaking from a porous water bucket. In a similar manner, leakage current is due to the imperfect quality of a capacitor's dielectric material.

The leakage resistance specification indicates the equivalent resistance of an external high value resistor connected across the capacitor, providing a high resistance discharge path for the charged capacitor. (See Figure 10.14)

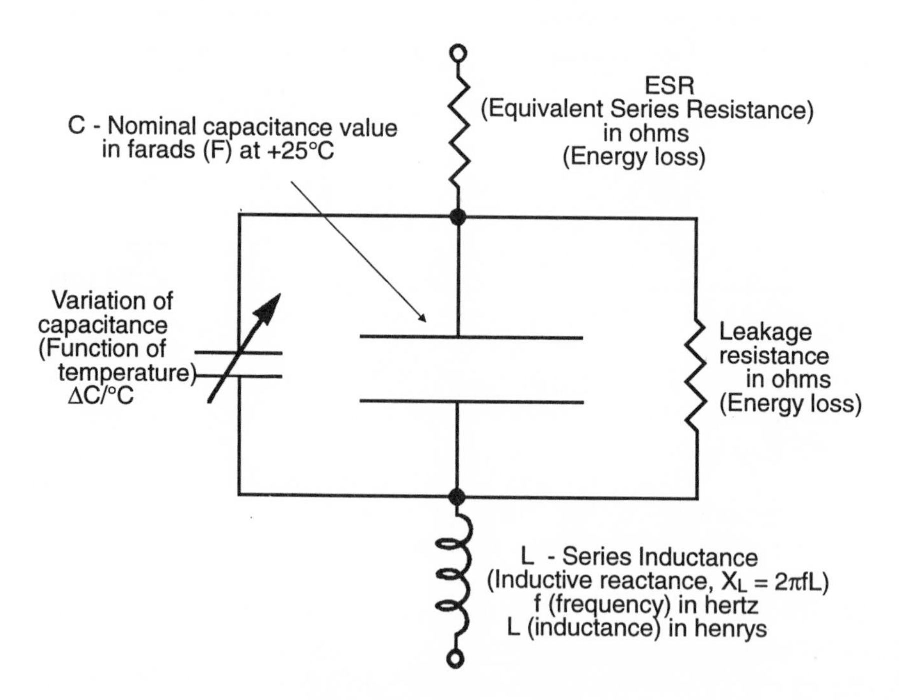

Capacitor Equivalent Circuit
Figure 10.14

SERIES INDUCTANCE (L) - Inherent in the design of some capacitors, a small, but undesireable characteristic called *inductance,* is produced as part of the capacitor and is measured in *henrys*. In the equivalent circuit of Figure 10.14, an inductance, L, shown in series with the capacitor, C, has a frequency-dependent AC resistance that increases as frequency increases.

This AC resistance, called *inductive reactance* , is designated as X_L, and is equal to $2\pi fL$, measured in ohms.

At high frequencies, X_L may be large enough to reduce charging current through the capacitor, negating its use at high frequencies. These types of capacitors should only be used at low frequencies.

FREQUENCY CAPABILITY - A chart of frequency capability versus capacitor type is shown in Figure 10.15. It represents the frequency capability or relative molecular mobility of different capacitor dielectric materials, as well as its inductive characteristics.

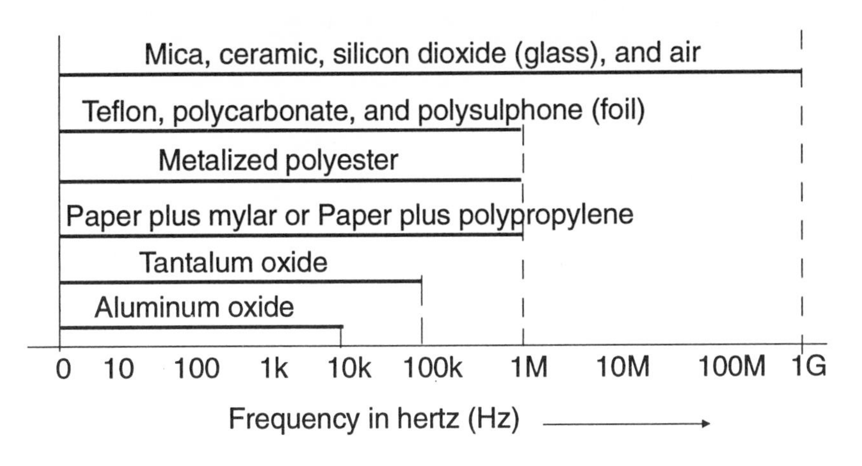

Frequency Capability as a function of Capacitor Type
Figure 10.15

PHYSICAL CHARACTERISTICS - Capacitors are made in a wide variety of package types and physical dimensions. These specifications include the capacitor's physical size, lead length and diameter, lead material, and mounting features, if applicable. These dimensions will often be shown in a manufacturer's data sheet or a customer's specifications drawing. Two typical types of lead arrangement are shown in Figures 10.16 and 10.17.

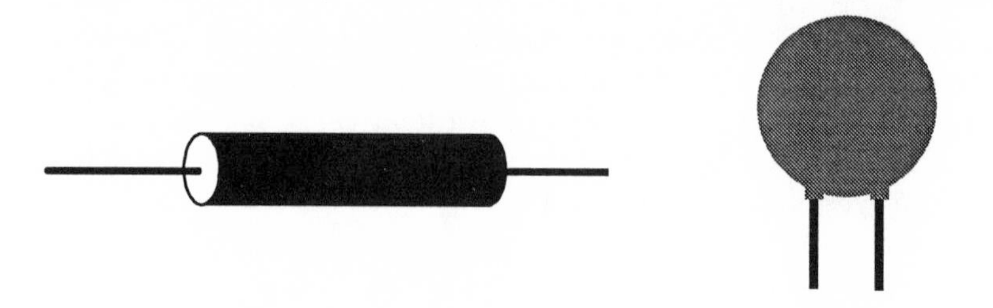

Axial-leaded Capacitor
Figure 10.16

Radial-leaded Capacitor
Figure 10.17

The diameter and length/thickness of a capacitor is called its *form factor* and is determined on the design of the capacitor, its manufacturing process, its dielectric, its capacitance, and voltage rating at +25°C. If another capacitor is made in the same type of package, with the same dielectric and the same voltage rating, but larger in capacitance, the form factor will also be larger. Generally, if a larger capacitance is desired, and the same form factor maintained, a capacitor with a lower voltage rating must be used.

For example, a metalized polyester capacitor specified by the manufacturer at 1 μF and 50 volts has a form factor (diameter x length) of 0.332 cubic inches. The same type of capacitor specified at 1 μF and 100 volts has a form factor of 0.537 cubic inches.

CAPACITOR CHIPS FOR HYBRID CIRCUITS

A **hybrid circuit** is an integrated circuit (IC) consisting of many discrete components in packaged form or unpackaged (chip) form. The components are assembled and soldered to a pre-wired single-layer or multi-layered *substrate.*

The substrate (also called passive substrate) is a pre-wired nonconductive material that provides the supporting or mounting surface for packaged components or for components in chip form. These components are assembled on the substrate to create a circuit, several circuits, a system, or an array.

The whole assembly is then enclosed in a package with terminals to be connected to an external system. The completed hybrid circuit is considered to be a single component or module.

The features of hybrid circuitry become apparent with the recognition that capacitors (and other components) in chip form have the same electrical characteristics as their packaged, discrete equivalents. In a hybrid circuit, all components are interconnected in a single module, packaged to offer the following advantages:

- Less handling in final assembly - ease in automatic insertion
- Reduced on-board space, over-all weight, and assembly costs
- Closer spacing between components to improve circuit performance and stability in space-critical applications
- Improved temperature tracking in the circuit
- Improved reliability because of reduced human handling

CAPACITOR CHIP TYPES

CERAMIC

- Connected layers of ceramic and aluminum in a monolithic structure
- Capacitance range: 1.0 pF to 1.5 µF and ± 1% to ± 2% tolerances
  Voltage range: 25 to 100 volt ratings
- Temperature coefficient: .003% to .025% $\Delta C/\Delta °C$
  Operating and storage temperature range: -55°C to +125°C

TANTALUM OXIDE

- Molded construction
- Capacitance range: 0.1 µF to 200 µF and ± 5% to 20% tolerances
  Voltage range: 4 volts to 50 volts
- Temperature coefficient:
  ± 10% change in capacitance from -55°C to +85°C
  ± 15% change in capacitance from +85°C to +125°C
- Operating and storage temperature range: -55°C to +125°C

SILICON

- Processed by semiconductor technology techniques
- Capacitance range: 1.0 pF to 100 pF and ± 1% to 5% tolerances
  Voltage capability: 50 volts
- Temperature coefficient: .015% to .035% Δ C/Δ°C
- Operating and storage temperature range: -55°C to +150°C

COMMON CAUSES OF CAPACITOR FAILURE

MANUFACTURING PROBLEMS

- Seal failure - The effects of humidity, fungus, salt spray, and other corrosive elements can cause degeneration of the capacitor and eventually result in its destruction because of a poor seal between the leads and the capacitor body.
- Residual dielectric contaminants can result in an increase in leakage or in the reduction the capacitor's life.

USER-GENERATED ERRORS

- Over-voltage or excessive surge current can result in dielectric rupture, destroying the capacitor.
- Incorrect polarity of DC voltage or AC applied to a polarized capacitor can result in its damage or destruction.
- Exposure to temperatures beyond the range specified in the data sheet can result in reduced voltage capability and possible capacitor destruction.

Capacitor production is a highly-developed and mature technology. As long as their specified ratings are not exceeded, capacitors will seldom fail.

REINFORCEMENT EXERCISE

Answer TRUE or FALSE

1. Capacitors are used for DC and AC filters, AC coupling, voltage sensing, timing, tuning, and other voltage storage applications.

2. A capacitor's ability to withstand stresses, such as humidity, fungus, salt spray, and chemical solvents depends on its dielectric material and has no relationship to its construction.

3. Although one capacitor may have the same capacity and voltage rating as another, a polarized type is physically larger than a nonpolarized type.

4. Aluminum oxide and tantalum oxide are used as the dielectric in polarized capacitors.

5. Aluminum oxide capacitors are preferred over tantalum oxide types in military and space applications where high temperature, stability, and reliability are prime requirements.

6. A polarized capacitor can fail if subjected to a voltage having an incorrect polarity.

7. An AC voltage can be connected across a polarized capacitor as long as a DC voltage is also applied across the capacitor to offset the AC from zero, and the magnitude of the DC voltage is greater than the amplitude of the AC voltage.

8. Variable capacitors are mostly used for high frequency tuning applications with air generally used as the dielectric. Mylar or mica can also be used as a dielectric in trimming applications.

9. Low leakage current in a capacitor specification indicates that its dielectric has very few contaminants and very high resistance.

10. The frequency capability of a capacitor is dependent on the molecular mobility of its dielectric and the inherent series inductance introduced by its manufacturing design.

11. The charge time of a capacitor in a circuit is dependent on its capacitance (in farads) and the resistance (in ohms) in series with

the capacitor in a charging circuit. The product of these two factors is called an RC time-constant and is measured in seconds.

12. A capacitor is discharged instantantly when a high resistance load is placed across its terminals.

13. Six RC time-constants are required to essentially fully charge or to discharge a capacitor.

14. The AC resistance, or capacitive reactance (X_C) of a capacitor will increase as the frequency of the voltage applied to its terminals is increased.

15. The electrical characteristics of capacitors in chip form are similar to those of equivalent packaged discrete capacitors using the same dielectric material.

16. Capacitor chips offer the advantage of reduction of on-board space, closer spacing between components to improve circuit performance, and improved temperature tracking.

17. Capacitor failures can be caused by a defective seal, aging, and/or dielectric contamination that could occur during the manufacturing process.

18. User-generated causes of capacitor failure include: DC voltage of incorrect polarity applied to the terminals of a polarized capacitor, excess applied voltage, excess surge current, and/or exposure to temperatures beyond their specified limits.

Answers to this reinforcement exercise are on pages 277-278.

CHAPTER
ELEVEN

SWITCHES, KEYBOARDS, AND RELAYS

MECHANICAL SWITCHES
- ACTUATING TECHNIQUES
- SPECIFICATIONS

KEYBOARDS , KEYPADS, AND KEYSWITCHES

ELECTROMECHANICAL RELAYS
- DEFINITIONS AND SPECIFICATIONS
- SPECIAL RELAYS
- RELAY PROBLEMS

REINFORCEMENT EXERCISE

SWITCHES, KEYBOARDS, AND ELECTROMECHANICAL RELAYS

MECHANICAL SWITCHES

A **mechanical switch** is a device that closes or opens an electrical or electronic circuit or transfers voltage from one part of a circuit to another. Other than hand-operated types, switches also provide the ability to sense a variety of stimuli and react automatically to control or change the state of a circuit or system.

A switch assembly consists of:

- A **contact set** (or sets) that closes and opens one or more circuits and/or transfers voltage from one part of a circuit to another.

A *normally-open* (NO) contact set can close, or energize, a circuit that is normally in a non-operating state. When the contacts are closed, the circuit is complete. A switch may have one or several sets of closing contacts called *make* contacts.

A *normally- closed* (NC) contact set can open, or de-energize a circuit that is normally operating. When the contacts are open, the circuit opens and no current flows. A switch may have one or several sets of opening contacts called *break* contacts.

A *transfer* contact set can have a combination of normally-open (NO) and normally-closed (NC) contacts and can transfer voltage from one part of a circuit to another.

- An **actuator** that is the mechanism operating the contact set.

- **Insulation** - a material that acts to isolate the contacts from each other and from the actuator, and to isolate the contacts from the frame.

- **A frame** (structure) that holds the entire assembly together.

SWITCH ACTUATING TECHNIQUES

A mechanical switch is initially specified by its *actuating technique,* the manner in which it is energized. Mechanical switches use a variety of the following actuating techniques, and sometimes, combinations of these techniques:

Hand-operated (manual)	Position	Magnetic
Thermal	Flow	Air pressure
Water pressure	Inertia	Acceleration

HAND-OPERATED (MANUAL) SWITCHES

PUSH-BUTTON SWITCH - Actuated by the push of a button.

TOGGLE SWITCH - A switch with a lever actuator or with a mechanism that moves the contacts between two positions and has no intermediate position

ROCKER-ARM SWITCH - Uses a pivoted lever with two flat surfaces at a 45° angle, or less. It is actuated by pressure on either face.

KNIFE SWITCH - The movement of its actuating handle resembles the opening or closing of a pocket-knife blade.

ROTARY SWITCH - Actuated by the rotation of a shaft. This switch is usually manufactured with two to twelve contact positions and one to four contact decks.

BAT-HANDLE SWITCH - Variation of a toggle switch - the actuating lever resembles the handle of a baseball bat

SLIDE SWITCH - Operated by a sliding motion of its actuator

THUMBWHEEL SWITCH AND JOGWHEEL SWITCH - The thumbwheel switch is actuated by the operator's finger (generally the thumb) moving a notched-wheel control. It is assembled in either a vertical or horizontal multisectional arrangement. The jogwheel switch substitutes a lever or a pushbutton for the notched wheel. Both switches can be convenient and economical encoding devices with a variety of digital codes available in standard assemblies.

TOUCH (MEMBRANE) SWITCH - Actuated by a touch on its touchpad with no perceptible movement of the actuating mechanism

DIP SWITCH - A multisection switch assembled in a miniature dual in-line package (DIP) with either slide, toggle, or rocker-arm controls. This switch is designed to be mounted on a PC board.

KEY OPERATED SWITCH - Actuated by inserting and then turning a key in a mating lock. It is usually used to enable or disable security systems or similar circuits.

Almost all of these switches may or may not have a spring-loaded mechanism, called a *momentary feature,* that causes the switch to spring back when the actuating force is removed.

POSITION ACTUATED SWITCHES

LEVEL SWITCH - This normally-closed (NC) switch is actuated when the level of a liquid has risen to a preset height established by the position of the float. When the switch is actuated, it will open a circuit to stop the flow of liquid into the container. (See Figure 11.1)

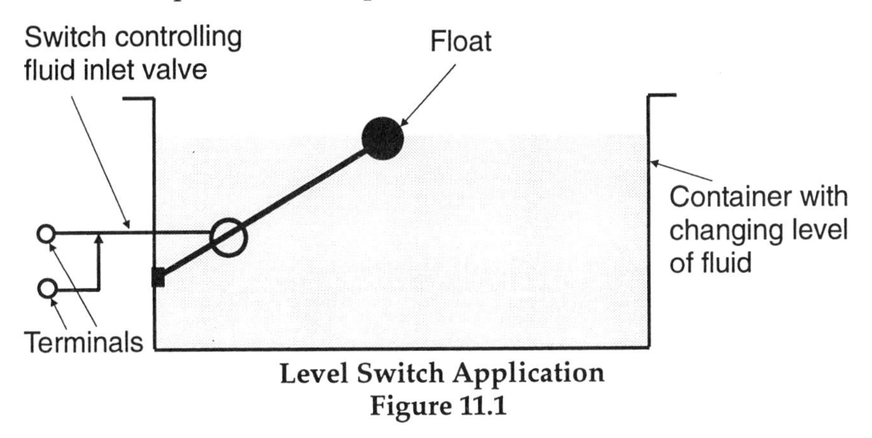

Level Switch Application
Figure 11.1

INTERLOCK SWITCH - Normally set in its actuated position (closed contacts) to maintain circuit closure. When the force that holds it in its actuated condition is removed, the spring-loaded switch will return to its open state, resulting in an open circuit.

A typical example of an interlock switch is the safety switch installed in the door of a microwave oven. When the door is opened and the force removed from the actuating mechanism of the switch, the DC voltage source for the microwave circuit will be removed. When the door is closed, the interlock switch will be set to its actuated state. Pushing the START button will close the microwave's DC voltage supply circuit and allow the oven to operate.

MOTOR-DRIVEN SWITCH - Actuated by the rotation of a bump or cam on the shaft of a timing motor when it comes in contact with the actuating mechanism of the switch.

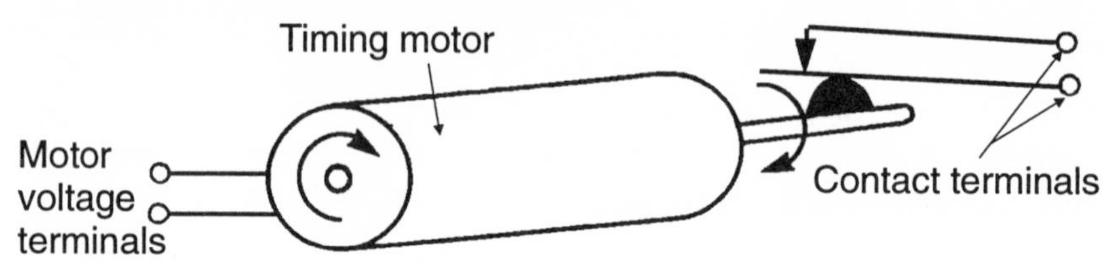

Motor-Driven Switch
Figure 11.2

MAGNETICALLY ACTUATED SWITCH

REED SWITCH - Actuated by the proximity of a magnetic field to its ferrous actuator or by the removal of a magnetic field from the vicinity of its actuator. The magnetic field is created by a movable, permanent magnet.

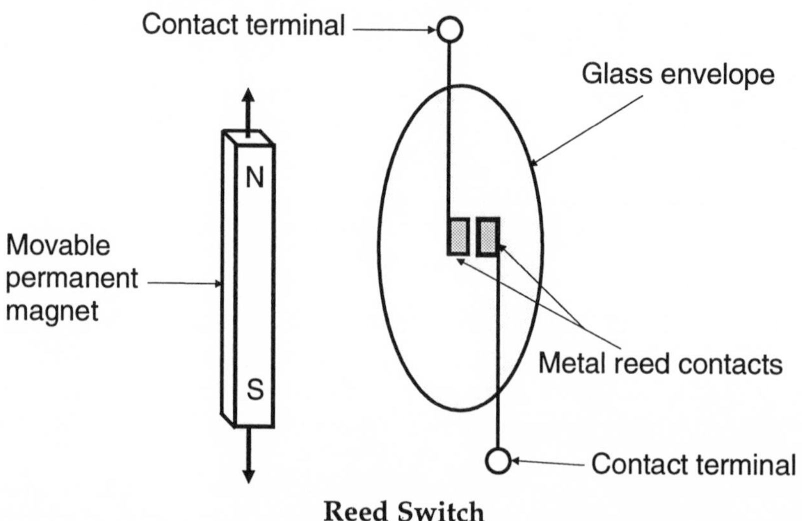

Reed Switch
Figure 11.3

OTHER ACTUATING TECHNIQUES

THERMAL SWITCH - This switch is actuated when exposed to a specified temperature. It is actuated in a snap-action manner at the rated temperature.

FLOW SWITCH - Actuated by a specified rate of flow of a liquid or gas along its sensing mechanism

PRESSURE SWITCH - Actuated when its sensing mechanism is subjected to a specified gas, liquid, or mechanical pressure

ACCELERATION SWITCH - This switch is actuated by applying or removing the force of acceleration at a specified rate. It is also called an INERTIA SWITCH.

MECHANICAL SWITCH SPECIFICATIONS

MAXIMUM CURRENT AND VOLTAGE - The maximum DC or AC current and voltage that can be safely handled under continuous operation during its expected life. The specification includes provision for a resistive or reactive load.

INSTANTANEOUS SURGE VOLTAGE AND CURRENT - Specifies the peak current and voltage that the switch contacts are capable of handling for a short duration, generally several milliseconds. Surge current is often referred to as *in-rush* current. It occurs at the moment of contact closure when the resistance of the load is lower on start-up than under normal operating conditions. A tungsten lamp is an example of this type of load.

INSULATION RESISTANCE - Indicates the quality of the insulating material. It is measured between each of the internal elements of the switch and between these elements and ground. Specified as a DC measurement, its value depends on the applied DC voltage.

DIELECTRIC STRENGTH - A measure of the insulating material's ability to withstand an AC voltage at a specified range of frequencies. Voltage is applied for one minute without breakdown or arc-over between the elements of the switch and frame. Other periods of time may be specified.

OPEN-CONTACT CAPACITANCE - When the metal contacts of a switch are physically in an open state, they act as the plates of an extremely small capacitor with air as its dielectric. At very high frequencies (RF), this small capacitance has a very low capacitive reactance (X_C) and acts to couple any high frequency signal from one contact of the open switch to the other contact. This capacitance, in picofarads, provides an undesirable, low resistance path for high frequency signals despite the open condition of the switch.

CONTACT BOUNCE - The normal action of a mechanical switch is to close or to operate when actuated. As the contacts meet, a mechanical force is produced that opposes the actuating force holding them together and pushes the contacts apart. Since the actuating mechanism is still exerting a force on the contacts to keep them together, they hit again, creating another opposing force strong enough to separate the contacts. This action can continue for several cycles until a stablized switch condition is reached. This transient condition of contacts opening and closing, called *contact bounce*, can create circuit errors and/or circuit instability.

CONTACT CONFIGURATION - The number and types of contacts of a switch are called its *configuration*. The moving contact is called the *pole*. A single pair of contacts is called a *single-pole* (SP). If there is only one option when the switch is actuated, to either close or open a circuit, that particular type of contact is called a *single-throw* (ST).

A single pair of contacts used to complete a circuit when actuated is designated as a *single-pole, single-throw normally-open* (SPST NO) switch. Unless otherwise indicated, a switch is shown on a schematic drawing in its non-actuated state. (See Figure 11.4a)

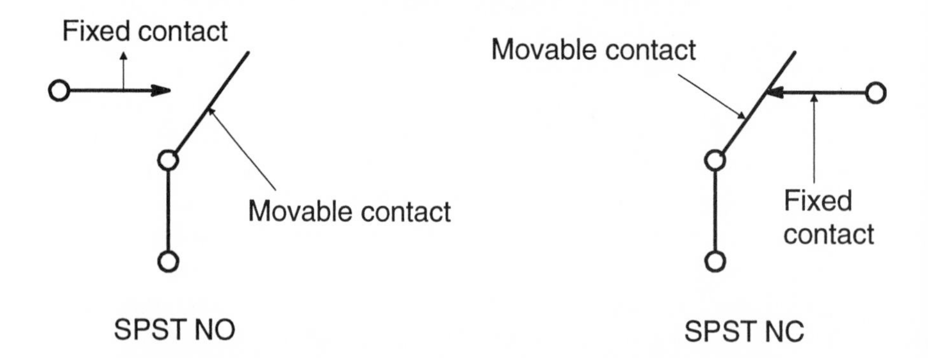

Single-pole, Single-throw Switch Configurations
Figure 11.4a **Figure 11.4b**

A *single-pole, single-throw, normally-closed* switch (SPST NC) is a single set of switch contacts that open a normally-closed circuit. (See Figure 11.4b)

A *make* is a set of normally-open switch contacts.
A *break* is a set of normally-closed switch contacts.

If the normally-open and normally-closed single-pole, single-throw switches are combined into a single switch with two switching options, a *transfer* capability results. With this configuration, voltage can be transferred from one point in a circuit to another when the switch is actuated. This is called a *single-pole, double-throw* switch (SPDT) having a set of transfer contacts. (See Figure 11.5a)

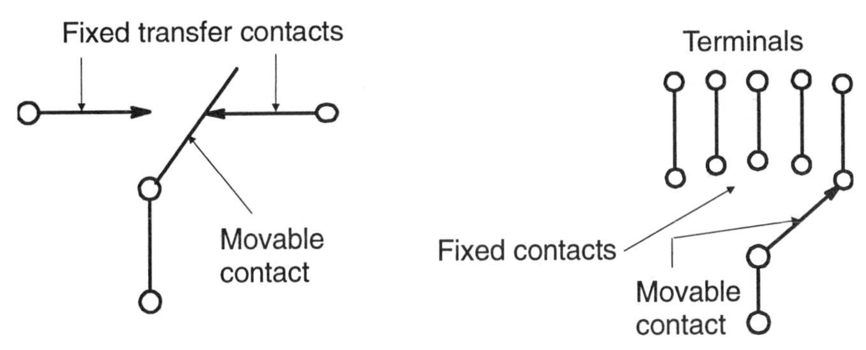

Single-pole Switch Configurations
Figure 11.5a **Figure 11.5b**

The SPDT switch can transfer in two ways:
- *Break-before-make* (non-bridging contacts) - break contacts are opened before the make contacts are closed.
- *Make-before-break* (bridging contacts) - break contacts stay closed until the make contacts are closed, then the break contacts are opened.

To transfer voltage from a point in a circuit to other optional positions, a *single pole, multiple position* switch is used. In Figure 11.5b, it is illustrated as a single-pole, 5-position (SP5P) configuration.

The switch can be designed for a break-before-make or a make-before-break configuration. Additional switching sections, or poles, can be added to actuate all the sections simultaneously to produce a configuration designated as *double-pole, triple-pole,* or *multiple-pole.*

The graphic symbols for double-throw, single-pole (DPDT) contact configurations are shown in Figures 11.6a through 11.6d.

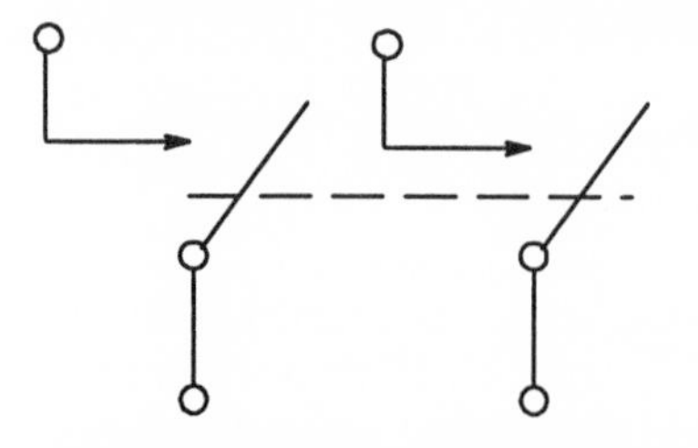

Double-pole, Single-throw Normally-open (DPST NO)
Figure 11.6a

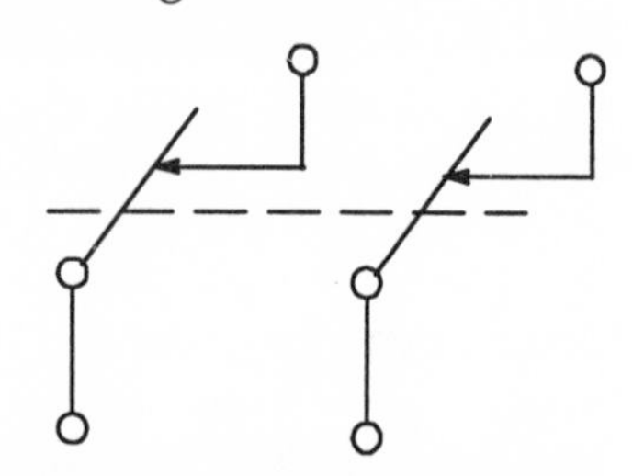

Double-pole, Single-throw Normally-closed (DPST NC)
Figure 11.6b

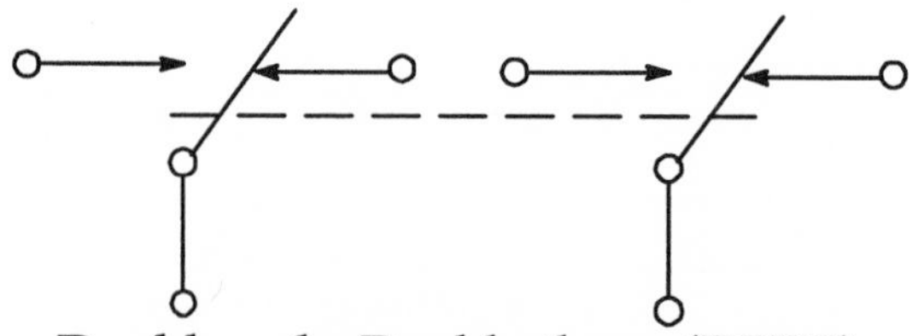

Double-pole, Double-throw (DPDT)
Figure 11.6c

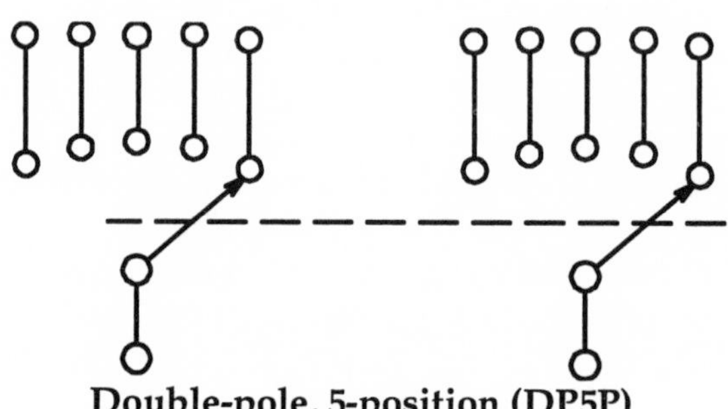

Double-pole, 5-position (DP5P)
Figure 11.6d

Note that the dotted line through the moving contacts of the switch indicates a *multiple-section* (ganged) switch. As one section is actuated, the other section is actuated simultaneously. There can be as many ganged sections, as needed, e.g.: 3-pole, single-throw; 4-pole, double-throw; 3-pole, 6-position; etc.

PHYSICAL CHARACTERISTICS

The physical characteristics of a device is supplied in a manufacturer's data sheet or in a customer's specification control drawing. The specifications information includes the following:

PACKAGE - Dimensions, housing material, and structural details of the switch

TERMINALS - Style, size, and material

CONTACTS - Provides details on the style of contacts, their size, and material. The resistance of a set of closed contacts is only a few milliohms, but this value may increase as the contacts become pitted or eroded because of excess current flowing through them. The end-of-life, or failure point contact-resistance depends on the application. If the current is high enough, an arc can develop across the contacts when they break, creating the possibility of the contacts corroding or welding. This type of failure can be avoided by the proper selection of contact size and material and by operating at a current level below the rated maximum current.

At very low values of voltage and current (50 mV or less at 1 mA or less), called a *dry circuit* condition, an oxide or sulfide film sometimes builds up on the surface of copper contacts. The oxide or sulfide film will increase contact-resistance and inhibit conduction at low voltage and current levels. To enhance conduction, it may be necessary to use other contact materials or to plate the surface of the contacts with silver, gold, other metals, or some appropriate alloy.

MOUNTING FEATURES - The required panel thickness and dimensions of the panel cutout, the style of mounting, or other mounting methods

ACTUATING FORCE - The minimum pressure needed to operate the switch

LENGTH OF STROKE - The minimum movement of the actuator necessary to operate the switch

ACCESSORIES - Normally supplied with the switch or offered as an option are accessories that include lighted caps, engraved caps, multi-colored caps, knobs, dial plates, bezels, and other miscellaneous mounting hardware.

OPERATING CAPABILITIES UNDER STRESS

OPERATING AND STORAGE TEMPERATURE - There is no standard operating temperature range for commercial, industrial, and consumer equipment. Typically, the temperature range extends from 0°C to +85°C. For military and space equipment, the standard operating temperature ranges from -55°C to +125°C.

RESISTANCE AGAINST ENVIRONMENTAL STRESSES - These stresses include shock, vibration, acceleration, humidity, fungus, salt spray, and chemical solvents.

OPERATING LIFE - A mechanical switch is subjected to friction, material fatigue, and normal mechanical wear. As a result, it has a limited life that is specified as the minimum number of operations it is capable of performing, or as the minimum number of hours of operation. One million operations is a typical value for a mechanical switch's operating life.

KEYBOARDS AND KEYPADS

KEYBOARDS

A **keyboard** is a data entry switch assembly resembling the traditional typewriter keyboard. Other keys are added that are used for control functions. The modern computer keyboard usually includes an interface logic circuit as part of its assembly which enables it to send encoded signals to digital systems.

Keyboards are used in a variety of applications, including computers, terminals, teletype, and other communications systems. Keyboards have been and are still being manufactured in a multitude of key arrangements and appearances. These configurations include: sloping or stepped-front panels, keyboards with or without sculptured buttons, a variety of color keys, key pressures, overtravel, key cap styles, and dimensions.

In an industry where custom keyboards have generally been used, there is now a trend toward standardization around the enhanced IBM PC keyboard. Several companies are manufacturing variations of this keyboard with interchangeability as a major feature.

KEYPADS

A **keypad** is an array of up to 20 keyswitches assembled in any preferred matrix (3 x 3, 3 x 4, etc.). It is manufactured either as a separate section on a keyboard panel adjacent to the keyboard, as a separate component in a system, or as a peripheral connected to a system.

Examples of keypad usage include: calculator data entry, microwave oven control, cursor control, numeric data entry, point-of-sale data entry, tape recorder control, and video game control.

Both keyboards and keypads are manufactured in full-travel and short-travel keyswitch technology, although full-travel keyboard switches are much more common.

FULL TRAVEL KEYSWITCH

When a keyswitch is depressed and travels about 0.12 to 0.15 inches, it is called a *full-travel* keyswitch. This distance is needed to achieve positive closure of the switch to generate the required encoded character. A spring or a compressed foam rubber pad provides a satisfying tactile "feel" for efficient typing.

The desired degree of key travel is very subjective and there are wide variations in the "feel" of the keys on different keyboards.

SHORT TRAVEL KEYSWITCH

When a keyswitch is depressed and travels less than 0.12 inch, it is called a *short-travel* keyswitch. This distance is needed to achieve switch closure to generate an encoded digital character. If a touch keyswitch is used, there is no perceptible movement of the key. In some cases, a tone, or click-sound, is produced to inform the operator that closure has been made.

Short-travel keyswitches are generally made with membrane technology. Short-travel keyswitches are more commonly used for keypads, however, they are sometimes used for keyboards as well.

KEYBOARD TYPES

Keyboards and keypads are classified according to the type of key switches used, such as: *mechanical* and *bounceless* keyswitches. The mechanical types are: the simple, hard, metal-to-metal contact type and the soft, metallized-plastic membrane type (*elastomer* or rubber).

Included in the bounceless category are Hall effect, capacitive, saturable ferrite core, and photo-optic switches. Bounceless keyswitches are generally more expensive, but provide trouble-free keyboard operation necessary for modern data entry equipment.

MECHANICAL KEYSWITCHES

SIMPLE MECHANICAL KEYSWITCH

If high speed data entry is not required, simple mechanical switches are specified for low cost consumer applications. They are generally included in smaller keyboards where the electronics section is an additional component of the equipment.

These keyboards have metal contacts that close when the key is depressed against a spring, actuating a circuit to generate the required encoded signal. On removal of finger pressure, the reaction of the spring resets the key to its non-depressed, open-circuit state.

With this switch, there is a possibility of generating multiple signals (contact bounce) when only one signal is desired. Protective circuits can be designed to avoid the effect of contact bounce, however, the cost of the electronics section is increased.

REED KEYSWITCH

The reed keyswitch is a variation of the purely mechanical keyswitch and is designed with a permanent magnet attached to a spring-loaded plunger. Depression of the key moves the magnet next to a glass-encapsulated reed switch to close its contacts. This action, in turn, closes a circuit to generate the required encoded signal.

A reed keyswitch offers excellent protection against humidity and other environmental contaminants; it is considered to be a long-life, highly reliable component.

MEMBRANE KEYSWITCH

A membrane keyswitch is a push-to-close-type, normally-open, momentary contact component. It is designed to switch low-energy DC, AC, and logic-level signals. As a completely sealed switch, it provides protection against liquids, dust, and other environmental contaminants.

A typical membrane keyswitch consists of a top flexible layer (elastomer) with conductive circuitry or shorting pads printed on its underside. A second stable layer has shorting pads printed on its top side. A spacer layer between the two conductive layers prevents them from coming in contact with each other until the flexible layer is depressed. Construction of a typical nontactile, conventional, membrane keyswitch in shown in Figure 11.7.

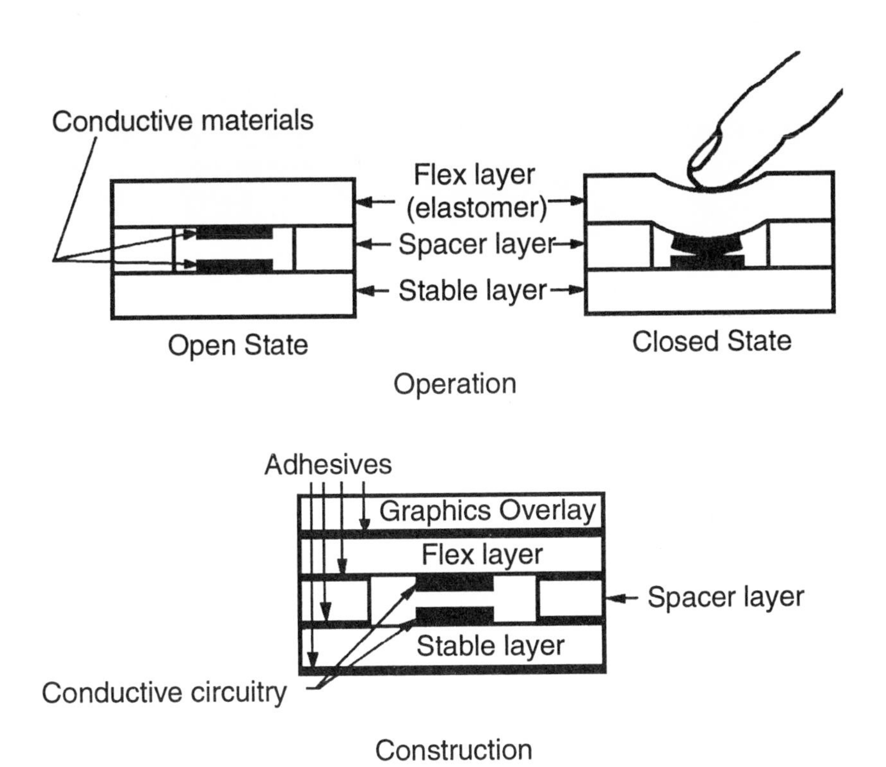

Conventional Membrane Keyswitch
Figure 11.7

All layers of the switch are held together by adhesives, including the decorative graphic display on the top of the flexible layer. When the key is depressed, the conductive sections of both top and bottom layers will come in contact with each other, actuating a circuit to generate the required encoded pulses. The drawings in Figure 11.8 illustrate the plastic dome, metal dome, and conductive elastomer tactile-type membrane keyswitches.

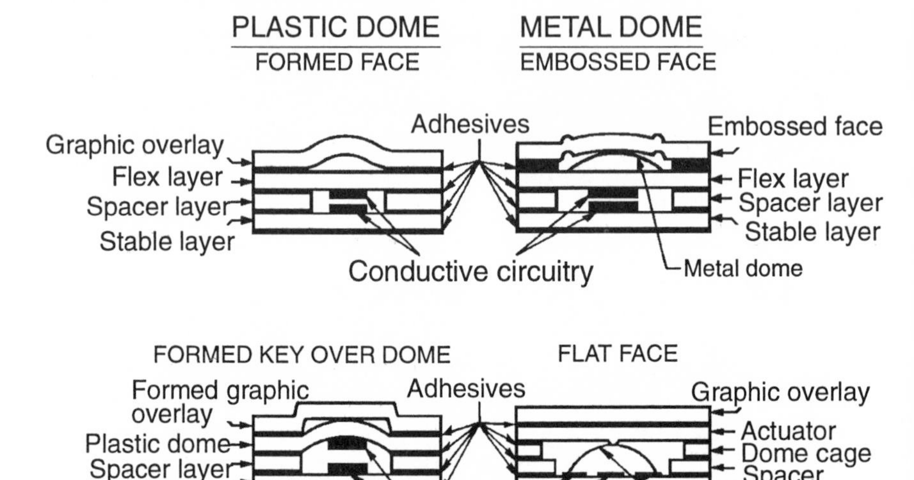

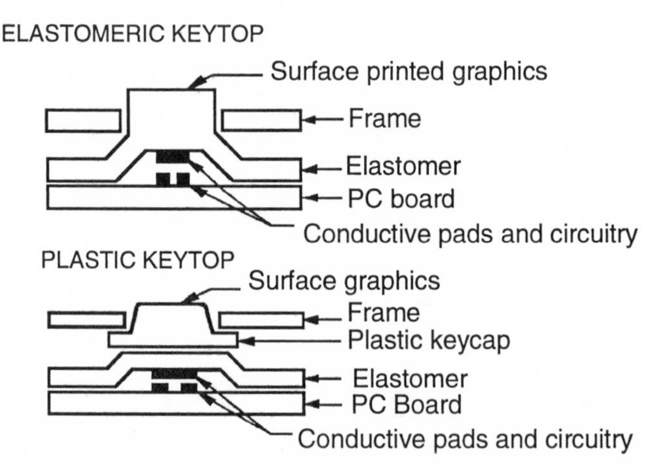

Construction of Various Tactile-Type Membrane Keyswitches
Figure 11.8

MEMBRANE KEYSWITCH CONSTRUCTION

GRAPHICS OVERLAY OPTIONS

- Can be made of polyester, polycarbonate, or be completely eliminated in favor of a mechanical actuator with no graphics
- Formed to provide a tactile feel or embossed to provide finger locations
- Textured or colored in an unlimited range of shades
- Coated with a variety of materials to protect against solvents and abrasions

FLEXIBLE LAYER
Typically made of polyester - can have conductive circuitry or shorting pads printed on its underside

SPACER LAYER (SPACER)
Thickness or cut-out sizes can be varied to change the pressure needed for actuation; a spacer is typically made of polyester or polycarbonate.

STABLE LAYER
Made of polyester or a single-sided printed circuit board.

ADHESIVES
Types will vary depending on the materials being laminated. Acrylic adhesive (the type generally used) affords long life, resistance to solvents, and operation at high temperatures.

CONDUCTIVE MATERIAL
Types will vary depending on the resistance levels and the environmental stresses applied to the switch.

MISCELLANEOUS CONSIDERATIONS
Construction will vary, depending on the need for:

- Additional layers to support the domes or other tactile forms
- Electrostatic shielding between face and flexible layer
- Special circuit layouts
- Special protection against environmental stresses

MEMBRANE KEYSWITCH SPECIFICATIONS

ELECTRICAL

CONTACT RATINGS (MAXIMUM) -
Typical values:
30 volts DC; 100 mA; 1 watt

CONTACT RESISTANCE - Varies proportionately with the size of the switch and conductive material used. For silver-based material, 100 ohms or less is typical; for PC boards, 1 ohm or less is typical.

MECHANICAL

OPERATING FORCE -
Typical values:
Nontactile type - 2 to 8 ounces
Tactile type - 8 to 16 ounces

SWITCH TRAVEL -
Typical values:
Nontactile type -.006" to .008 "
Tactile type -
Metal dome - 0.2" to 0.3"
Plastic dome - 0.25" to 0.35"
Elastomeric type - Typical values - .01" to 0.1"

CONTACT BOUNCE -
Typical values:
1 millisecond nominal
10 milliseconds maximum

ENVIRONMENTAL

OPERATING AND STORAGE TEMPERATURE RANGES
Nonvented range: -40°C to +65°C
Externally vented range: -40°C to +85°C

OPERATING LIFE
Typical values:
5 million to 10 million cycles

BOUNCELESS KEYSWITCHES

HALL EFFECT KEYSWITCH

The *Hall effect*, developed in 1880 by **Edwin H. Hall**, an American physicist, is used to produce a Hall effect keyswitch that incorporates an IC chip in its structure containing a generator, a trigger circuit, and an amplifier. When the key is pressed, a permanent magnet moves across the IC chip placing a magnetic field around it. The magnetic field produces a voltage at the output of the Hall generator switching the trigger circuit to its ON state. The output of the trigger circuit is an encoded signal representing the specific character assigned to the key being pressed. (See Figure 11.9)

Hall effect keyboards provide bounceless switching, are relatively insensitive to hostile environments, and provide excellent reliability. They are capable of a minimum of one million switching operations.

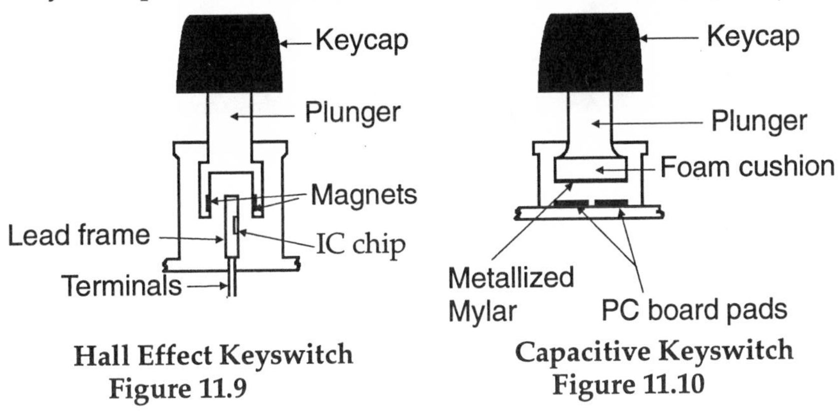

Hall Effect Keyswitch
Figure 11.9

Capacitive Keyswitch
Figure 11.10

CAPACITIVE KEYSWITCH

Capacitive keyboards are capable of more than 1,000,000 operations and provide virtually noise-free switching at low power. In a capacitive switch, two pads on the printed circuit (PC) board under the key act as capacitor plates connected to the drive and sensing circuits. (See Figure 11.10)

Pressing the key causes an increase in the series capacitance, enhancing the coupling between the pads on the PC board and producing a voltage in the sensing circuit. By using large pads, a high ON/OFF capacitance ratio is achieved assuring consistent switch operation.

SATURABLE FERRITE-CORE KEYSWITCH

A saturable ferrite-core keyswitch is capable of performing more than one million operations with very little, if any, sensitivity to electrostatic discharge. It is considered to be a high reliability switch with a satisfying keyboard "feel".

- In the OFF position, permanent magnets saturate a ferrite-core and prevent a 500 kilohertz drive signal from being picked up by the sensing wire through transformer coupling (induction). (See Chapter Twelve - Magnetic Components for a more detailed discussion of transformer action and induction.)
- When the key is pressed, the magnetic field is removed allowing the drive voltage to be coupled to the sensing wire. The switch turns ON, closing the circuit that generates the appropriate encoded signal. (See Figure 11.11)

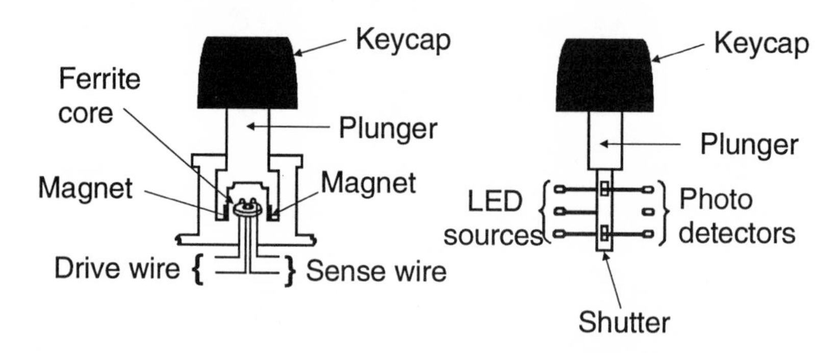

Saturable Ferrite-Core Keyswitch
Figure 11.11

Photo-optic Keyswitch
Figure 11.12

PHOTO-OPTIC KEYSWITCH

In photo-optic keyboards, the electronics section contains detector and encoder circuits for each key. Keys may easily be added, deleted, or exchanged, offering the flexiblity for developing additional prototype keyboards.

In the photo-optic keyboard, beams of light are directed through a key matrix onto photo-detectors. Each keyswitch has a notched steel member (shutter) that, when pressed, selectively interrupts the light beams. The output of the photo-detectors connect to amplifiers that generate the proper encoded signal. (See Figure 11.12)

ELECTROMECHANICAL RELAYS

An **electromechanical relay** operates as a switch, or as a group of ganged switches, actuated by the application of a control voltage applied to the terminals of a magnetic coil.

When voltage is applied to the coil terminals of a relay, the resulting current through the coil creates a magnetic field around the coil and around the elements of the relay, and acts as an electromagnet. Its magnetic field attracts a ferrous, spring-loaded, moving pole, called the *armature*, thereby operating the switch contacts of the relay.

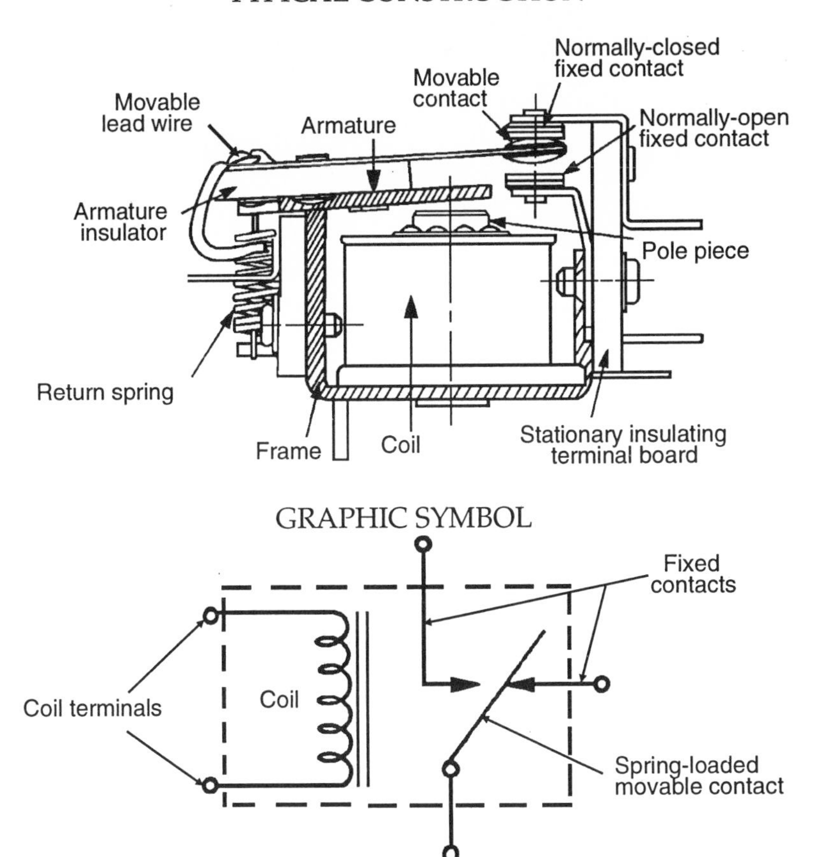

Electromechanical Relay
Figure 11.13

ELECTROMECHANICAL RELAY FEATURES

- Since the actuating voltage of a relay can be applied at a distance from the switching section, it has remote control capability.
- Modification of the coil circuit can provide time-delay capabilities for operating and releasing the relay.
- With the control voltage being supplied from an electrical or electronic circuit, the need for a mechanical force or manual movement to turn a switch ON or OFF is eliminated.

ELECTROMECHANICAL RELAY SPECIFICATIONS

To a great extent, the specifications for an electromechanical relay are similar to the those of a mechanical switch with the added specifications of coil voltage and current.

COIL RATINGS

COIL VOLTAGE -
A coil is specifically wound to operate with either AC and/or DC.
- DC - Nominal values range from 3 to 115 volts
- AC - Nominal values range from 6 to 220 volts

OPERATE AND RELEASE COIL CURRENT
- Pull-up (operate) current - Minimum current level at which the relay will operate
- Drop-out (release) current - Maximum current level at which the relay will return to its nonactuated state

DC RESISTANCE OF THE COIL
Specified in ohms

INDUCTANCE
An inherent characteristic of a coil, specified in *henrys*. Inductance is a function of the number of turns in the coil and the type of core material used in winding the coil. When AC voltage is applied to the coil, an AC resistance, called *inductive reactance* , is created that acts to limit the current in the coil.

OPERATE AND RELEASE TIMES

- Operate time - Minimum interval between the time coil voltage is applied and the time that actuation of the contacts occurs

- Release time - Maximum interval between the time the coil voltage is removed and the time the contacts return to their non-actuated state

INSULATION RESISTANCE
The quality of the insulating material between coil and frame. It is a DC measurement that depends on the value of applied DC voltage.

DIELECTRIC STRENGTH
The measure of the insulating material's ability to withstand a specified AC voltage between coil and frame. It is defined as a maximum AC voltage at a specified frequency range which can be applied for one minute, or a specified period, before breakdown of the insulation (arc-over) occurs.

CONTACT RATINGS

VOLTAGE AND CURRENT - Maximum values of AC and/or DC voltage and current that can be safely handled by the relay contacts.

INSULATION RESISTANCE - The quality of the insulating material between the contacts and between the contacts and frame. Insulation resistance depends on the magnitude of the DC voltage.

DIELECTRIC STRENGTH - Defines the insulating material's ability to withstand a specified AC voltage between the contacts and between the contacts and frame. It is measured as a maximum AC voltage at a specified frequency range, applied for one minute or for a specified period, before breakdown of the insulation occurs.

CONTACT CAPACITANCE - With the contacts in an open state, its metal sections create a very small capacitor with air as its dielectric. At very high frequencies, the capacitive reactance (X_C) is very small, thereby effectively coupling the signal from one contact of the open switch to the other. With the switch contacts open, the low AC resistance path across the contacts is unacceptable since the resistance between them should be infinite.

CONTACT CONFIGURATION

Drawn in its nonactuated state and designated as:

- Form A - Make - normally open (NO)
- Form B - Break - normally closed (NC)
- Form C - Transfer

The *form* designations, A, B, and C, are used for ease in data sheet specifications (see Figure 11.14) with the form letter preceded by the number of contact sets in that particular group.

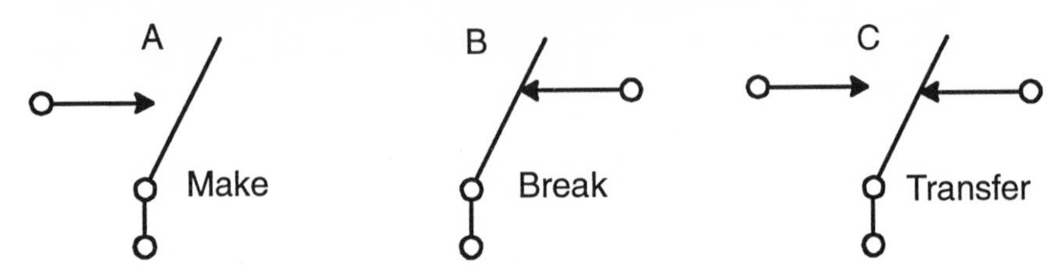

Types of Contact Form
Figure 11.14

For example, typical relay configurations can be specified as follows:

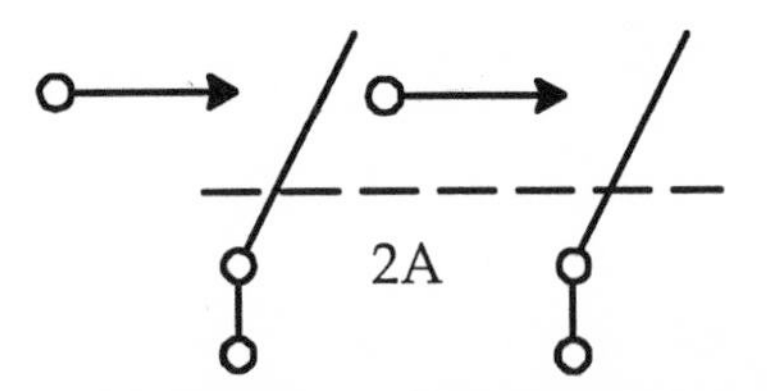

Contact form 2A designates two sets of make (A) contacts

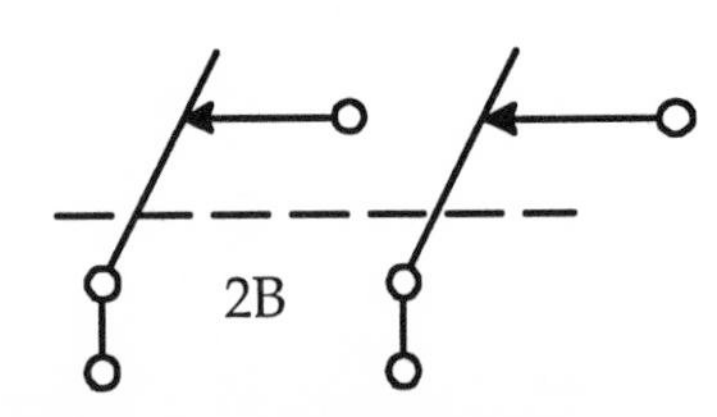

Contact form 2B designates two sets of break (B) contacts

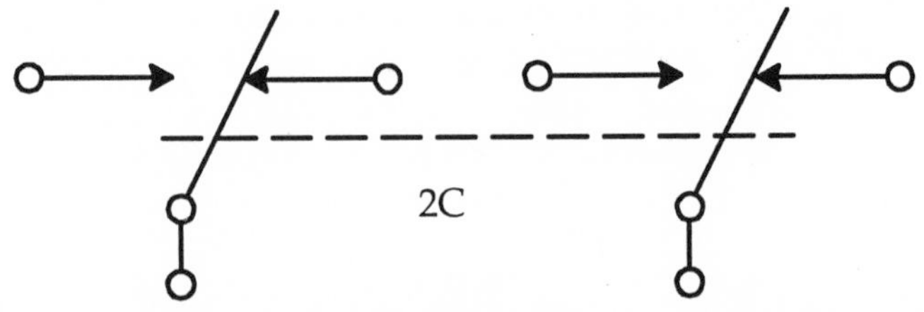

Contact form 2C designates two sets of transfer (C) contacts

Figure 11.15

Letter designations of basic relay contact configurations and their variations have been standardized by the National Association of Relay Manufacturers. (See Figure 11.16)

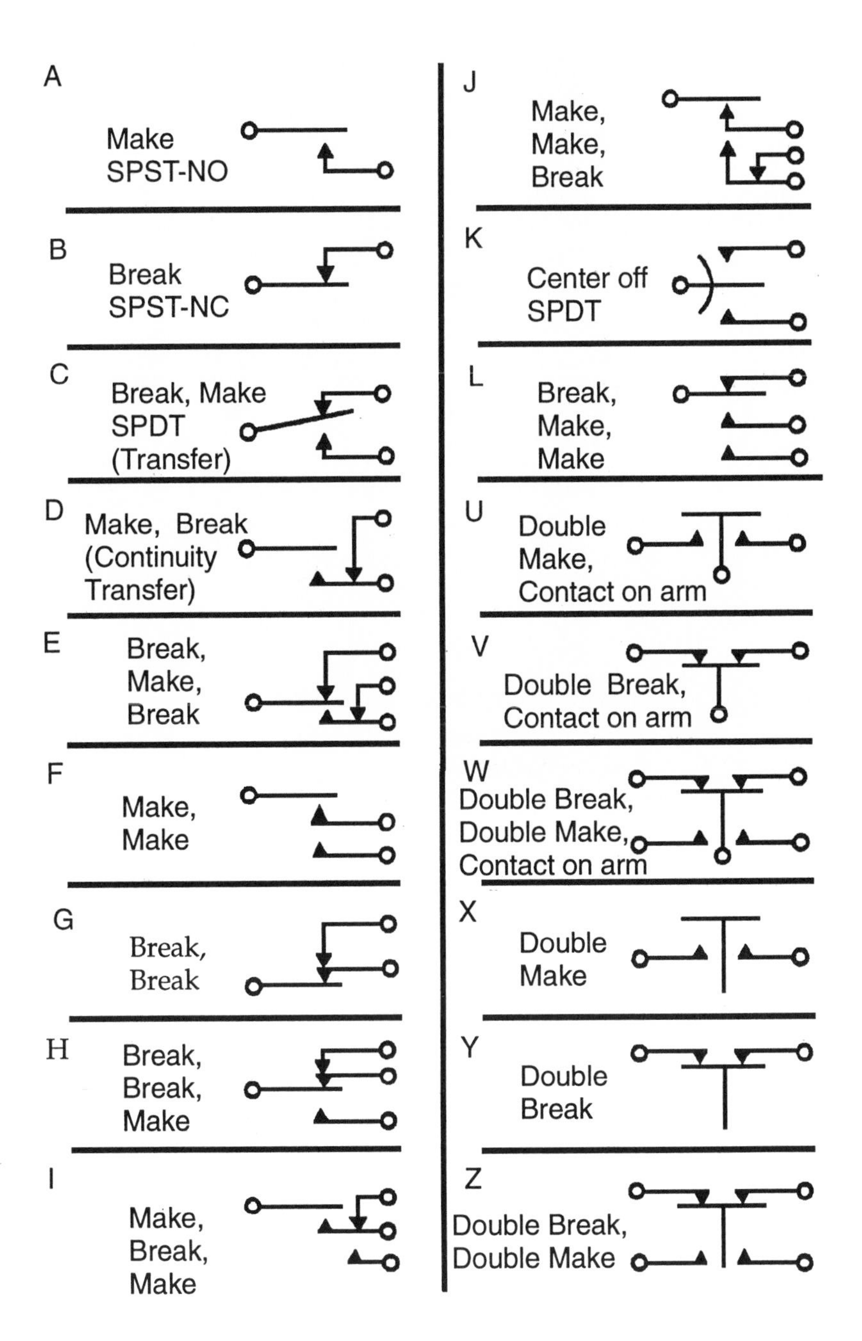

Note: All contacts assume a downward movement of the armature upon actuation

Standard Relay Contact Forms
Figure 11.16

PHYSICAL SPECIFICATIONS

PACKAGE
Specified by dimensions, weight, material and structure. Types and housings exists that include telephone styles for communications and open frame types for general purpose and high current applications. Miniaturized packages for low current, low voltage applications include plastic dual in-line packages (DIPs) for commercial, industrial, and consumer equipment. Hermetically sealed crystal-can packages and transistor-type metal cans (T0-5) are intended for military and space applications.

MOUNTING FEATURES
A variety of relay layout, positioning, and mounting are specified, including: rack, panel, chassis, printed circuit board, socket, and a variety of other types

TERMINALS
Specified according to type, size, material, and spacing between terminals for PC board mounting

CONTACTS
Details on the type, size, and material of contacts are covered on page 197 in the Switch section of this chapter.

ENVIRONMENTALCONSIDERATIONS

OPERATING AND STORAGE TEMPERATURE RANGE
For commercial, industrial, and consumer applications, no standard range of temperature exists. The temperature range depends on the intended application. Typical temperature for these applications ranges from 0°C to +70°C.

The temperature range for military and space equipment is from -55°C to +125°C.

RESISTANCE AGAINST ENVIRONMENTAL STRESS - Where applicable, these values are specified for shock, vibration, acceleration, humidity, fungus, salt spray, and chemical solvents.

OPERATING LIFE - Rated life is specified as a minimum number of operations or as a minimum number of hours of operation.

SPECIAL RELAYS

Standard electromechanical relays are available in all sizes, configurations, operating characteristics, mounting techniques, and applications. In addition, there are electromechanical relays manufactured for special uses, improved performance characteristics, and enhancement of relay switching capability.

SOLENOID

A *solenoid* consists of a coil into which a movable, spring-loaded iron or steel core, called a *plunger* or *armature,* is assembled. When a DC voltage is applied across the terminals of the coil, an electromagnet is created, causing the plunger to move within and along the axis of the coil for a specified distance. A solenoid converts electrical energy (the voltage and current applied to the coil) into linear mechanical movement to actuate a variety of electrical, mechanical, or pneumatic systems.

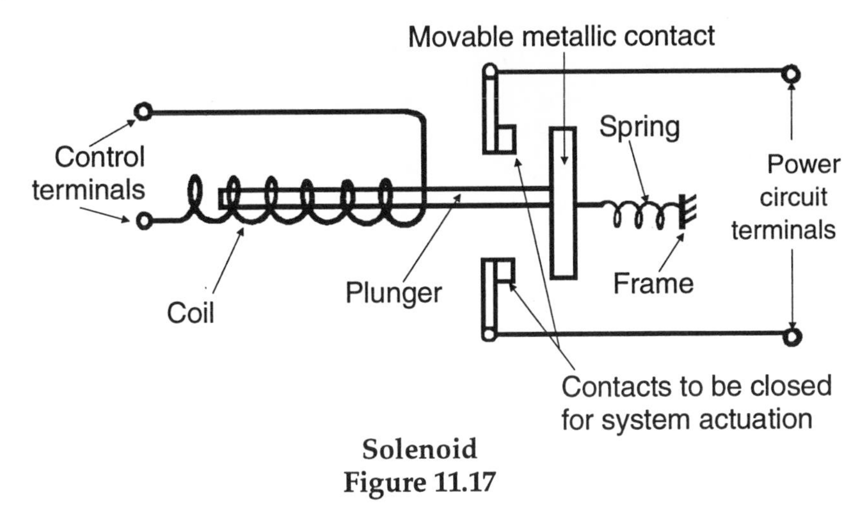

Solenoid
Figure 11.17

The plunger is spring-loaded during the solenoid's nonactuated state. When the solenoid coil is actuated by applying a voltage across the coil, a magnetic force is applied to the plunger to overcome the force of the spring and deliver the desired movement. When the coil voltage is removed, the plunger will return to its original position, de-energizing the electrical, mechanical, or pneumatic system.

The coil of the solenoid requires sufficient electrical power to pull the plunger the required distance. The number of turns in the coil and the magnitude of the current in the coil provide the necessary magnetic field for sufficient plunger movement.

REED RELAY

Standard electromechanical relays respond to actuating voltages in the order of hundreds of milliseconds. In modern electronic circuitry, this timing may be too slow and counter-productive to the needs of a system. Operating times can be shortened by using a *reed relay*, a relatively small, sensitive electromechanical relay.

A reed relay consists of a coil wound around a sealed glass capsule containing a single set, or multiple sets, of magnetically-sensitive, metallic reeds. The entire assembly is then enclosed in an appropriate package, typically a DIP. They are, however, available in a wide variety of packages. Rhodium relay contacts are mounted on the end of the reeds which contact each other when the coil is energized, closing the switching circuit. On the removal of coil voltage, the reeds move apart, opening the switching circuit.

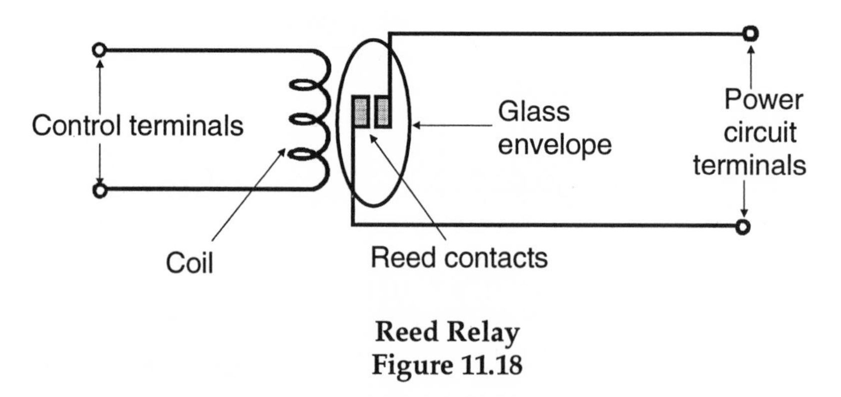

Reed Relay
Figure 11.18

Reed relays typically have operate and release times of 300 to 1000 microseconds and are capable of performing over a billion operations (about a thousand times more than standard electromechanical relays). Coil voltage requirements range from 5 to 24 volts DC with coil current levels generally lower than 5 to 30 milliamperes. Since a reed relay uses very low operating coil current, it is sensitive to any extraneous, or stray, magnetic fields that might be in the vicinity of the relay. Some shielding against these magnetic fields may be required to prevent inadvertent activation (or operation).

Typically, these relays can switch resistive loads up to a maximum of 500 milliamperes and up to a maximum DC voltage of 200 volts. Greater longevity can be achieved, however, if the contacts are operated at lower than maximum current levels.

MERCURY-WETTED RELAY
This relay contains a pool of mercury at the base of the contact armature to overcomes the effects of contact bounce described on page 216. Through capillary action, mercury flows from the base to the contacts to wet the surface of the contacts. When the relay is actuated, the wetted contacts maintain a mercury-to-mercury connection (electrical closure), despite any bouncing of the contacts.

Although a mercury-wetted relay switches more slowly than a reed relay, it is faster than the standard electromechanical type. The mercury film virtually eliminates corrosion of the contacts, however, it is relatively sensitive to shock and vibration and is generally mounted in a vertical or near vertical position. Since mercury freezes at -38.8°C, operation at extremely low temperature is limited.

STEPPER RELAY
Moves a rotary switch through a precise angle by each input pulse voltage applied to its coil. The opening and closing contacts are operated by ratchets or similar mechanisms at each different angular position. A stepper relay can be electrically or mechanically reset to its initial or home position.

LATCHING RELAY
Maintains its actuated state despite the removal of the actuating coil voltage. Resetting the relay to its nonactuated state can be accomplished either manually or electrically.

RF RELAY
Designed for switching radio signal currents with a minimum attenuation of the RF signal in its ON state and provides essentially infinite resistance in its OFF state.

TIME-DELAY RELAY
Provides specific operate and/or release times that are set by the relay manufacturer. Adjustable time-delays can be achieved magnetically, mechanically, or thermally, or with an RC time-constant in the coil circuit. Adjustment of a variable time-delay relay is generally an end-user function.

VACUUM RELAY
Designed with contacts that are sealed in a glass-enclosed vacuum to prevent contamination in an environment of dust or other impurities. A vacuum relay has the capability of switching high currents and high voltage.

ELECTROMECHANICAL RELAY PROBLEMS

CONTACT BOUNCE - Similar to the switch action described on page 194, the normal action of an electromechanical relay is such that its "make" contacts hit each other suddenly, producing a mechanical force that opposes the magnetic force pulling them together. This opposing mechanical force tends to push the contacts apart to open the circuit. Since the magnetic field is still exerting a pulling force on the moving pole (the armature), the contacts hit again, creating another opposing mechanical force strong enough to separate the contacts again. The opposing forces produce bouncing of the contacts which can continue for several cycles until stability is reached and the relay is settled in its fully operating, or ON, state.

Contact bounce can create circuit errors and/or circuit instability and should be minimized or eliminated by proper component selection and/or appropriate circuit design. The time for this transient condition is specified in seconds (or milliseconds).

PITTING AND GENERATION OF RFI - Normal operation of an electromechanical relay causes abrupt closing and opening of its contacts. In a high power circuit, this normal switching operation creates a spark or arc (ionization or breakdown of the air) at the contacts of the relay. As the relay continues to operate in this manner, the continued generation of arcs can cause the contacts to become eroded (pitted), resulting in poor conductivity between contacts. In addition, arcing generates electromagnetic waves called *radio frequency interference* (RFI) that can cause disturbances in electronic equipment in the vicinity of the relay. It is often necessary to minimize or eliminate RFI to maintain proper circuit operation.

NORMAL MECHANICAL WEAR - With any mechanical system, there is a tendency for the system to wear down because of the effects of friction, material fatigue, and the natural forces of erosion and contamination. In the case of electromechanical relays, this inherent characteristic produces reduced life and eventual failure. Although relay life may be sufficiently long for practical purposes, this normal deterioration may preclude the selection of an electromechanical relay in favor of other types of switching devices for longer lasting, reliable circuit operation.

The solid-state relay (SSR), a semiconductor component that eliminates these problems, is covered in Volume Two - Part Two - Optoelectronics.

REINFORCEMENT EXERCISE

Answer TRUE or FALSE

1. A switch is specified by its actuating technique, its contact configuration, its electrical ratings (voltage, current and power capability), its physical characteristics, mounting features, and its ability to withstand environmental stresses.

2. A toggle switch is the general term for a snap-action switch and is available in a variety of types, such as: rocker-arm switch, slide switch, and bat-handle switch.

3. A DIP switch is a multi-section switch assembled in a dual-in-line package (DIP) with either slide or rocker-arm controls.

4. Actuation of a magnetic switch is achieved by the application of a voltage to the switch terminals.

5. Since switches are not involved with friction or wear, they have infinite life capability and are not limited to a specific number of operations.

6. Bounceless keyboards are preferred for use in computer equipment because their switchkeys are quieter than the mechanical types and provide a more pleasing tactile feel for the operator.

7. An electromechanical relay is actually a switch, or combination of switches, actuated by the application of a voltage at the terminals of a coil to produce a magnetic field around the coil. The magnetic field is used to attract a spring-loaded, moving contact (armature) to provide switching action.

8. The specifications of a relay are essentially the same as those of a switch with the added electrical specifications of the coil.

9. A relay can only be actuated by applying DC voltage to the coil.

10. A dry circuit is one that has an absence of moisture, allowing the circuit to operate under all conditions of voltage and current.

11. Contact configurations are drawn in their nonactuated state in a manufacturer's data sheet or a customer's specification sheet.

12. A reed relay is a small, sensitive switching device, enclosed in a protective glass capsule, operating at a much higher switching speed, and having a longer lifetime than an ordinary electromechanical relay.

13. A mercury-wetted relay provides improved conduction of its switching contacts by wetting its contacts with a mercury film.

14. Mercury-wetted relays must always be mounted in a horizontal position and never in a vertical or near vertical position.

15. The existence of contact bounce inherent in an electromechanical relay has no detrimental effect on the stability of a circuit and is not taken into account in the design of electronic systems.

16. The arcs that can be produced by switching an electromechanical relay ON and OFF, generate radio frequency interference (RFI). If a circuit is not properly protected against RFI by shielding and other techniques, these interferences can act to disturb the system operation.

Identify the contact configuration of each of the following symbols:

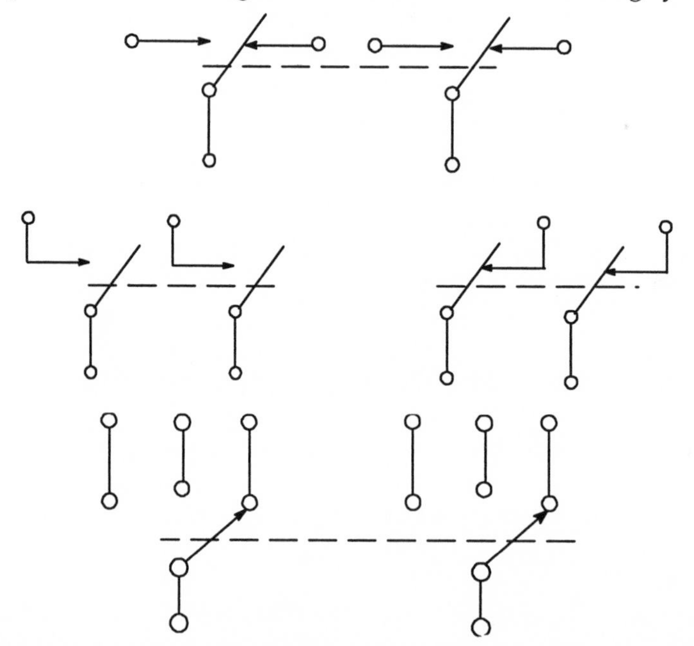

Answers to this reinforcement exercise are on pages 279-280.

CHAPTER
TWELVE

MAGNETIC COMPONENTS

MAGNETIC COMPONENTS

INDUCTORS

An **inductor** is a component consisting of wire that is wrapped around a ferrous (iron or steel) or non-ferrous core. The wire is coated with insulating material, such as varnish, enamel, or lacquer to prevent the turns of wire from making electrical contact with each other or with the core.

Inductor is the generic term for the component generally called *choke* or *coil*. In some applications, these terms are used interchangeably.

Inductors provide frequency-dependent resistance and are used in circuits with AC or pulsating DC voltages to emphasize or discriminate against specific frequencies.

INDUCTANCE

When a wire is wound around a core to form an inductor, a magnetic characteristic called *inductance* is created. This characteristic is a function of the number of turns of wire wound around the core and the type of material used as the core. The frequency of the applied voltage will determine the core material of the inductor.

- The greater the number of turns in an inductor, the higher the inductance.

- The more ferrous material in a core, the higher the inductance. Ferrous materials include: iron or steel laminations, or powdered iron (ferrite).

- When a core is a nonferrous material, there is still inductance in the component, although it is of a significantly lower value. This type of device is referred to as an *air core* inductor .

Inductance is designated with the letter, **L**, and its unit of measurement is the **henry (H)**, named after **Joseph Henry**, a 19th century American physicist.

CORE MATERIAL

Inductors are categorized according to their core material, as follows:

- Iron or steel laminations are used at operating frequencies that range up to 20 kilohertz.
- Ferrite or powdered iron is used at operating frequencies ranging from 20 kilohertz to about 10 megahertz.
- Nonferrous materials such as aluminum, plastic, and ceramic, or air are used at operating frequencies above 10 megahertz.

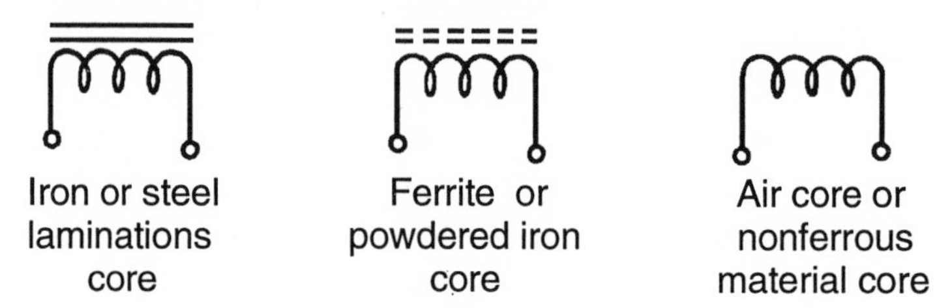

Graphic Symbols for Coils, Chokes, and Inductors
Figure 12.1

INDUCTIVE REACTANCE

When alternating current (AC) flows through a coil, an alternating magnetic field is produced aroundthe coil. This changing magnetic field produces a voltage, the *counterEMF*, opposite in polarity to the applied voltage. The counter EMF creates the *inductive reactance*, a *frequency-dependent* AC resistance that opposes the flow of current in the circuit. Inductive reactance is measured in ohms and symbolized as $\mathbf{X_L}$, a variable value that depends on:

- The specified value of its inductance, **L**, in henrys
- The frequency, **f**,, of the applied AC voltage, in hertz
- The arithmetic constant: 2π (3.14), with 2 x π = 6.28

Inductive Reactance = $\mathbf{X_L = 6.28 \times f \times L}$ in ohms

Inductive reactance is directly proportional to both the specified inductance of the coil and the frequency of the applied voltage.

AC AS THE SOURCE OF VOLTAGE

When an AC voltage is connected to an inductor (the load in the circuit of Figure 12.2), having negligible or no DC resistance, the circuit current (I_L) is calculated by first solving for the inductive reactance (X_L) of the load and then using this value to calculate I_L.

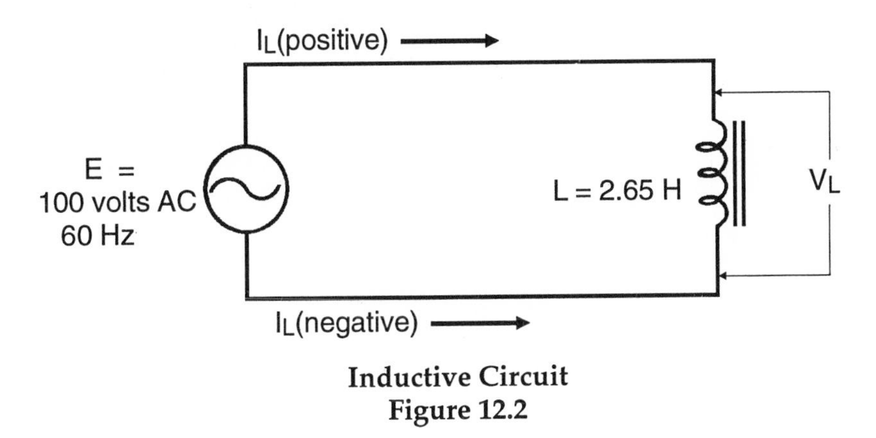

Inductive Circuit
Figure 12.2

If E = 100 volts AC at 60 Hz and the inductance, L, of the coil is 2.65 henrys, then:

$$X_L = 6.28 \times f \times L$$

$$X_L = 6.28 \times 60 \times 2.65 = 1000 \text{ ohms}$$

and

$$I_L = \frac{E}{X_L} = \frac{100}{1000} = 0.1 \text{ amperes or } 100 \text{ mA}$$

DC AS THE SOURCE OF VOLTAGE

When steady-state direct current flows through a straight (uncoiled) length of wire, a constant magnetic field is produced around the wire. The DC resistance is a fixed value that depends on the length, diameter, and molecular structure of the wire. If the same length of wire is wound around a core to form a coil, its DC resistance does not change from its original value.

Since the frequency of steady-state DC voltage is zero , the value of "f" in the inductive reactance equation is zero. When steady-state DC is used as the supply voltage, the inductive reactance is zero, regardless of the value of inductance. If the wire used to wind the coil has extremely low resistance (assumed to be zero), then the current in the circuit will be extremely high, possibly high enough to damage the circuit or blow a fuse.

In some applications where a magnetic field is desired and DC is used as the voltage source, a sufficiently high DC resistance can be achieved by adding a resistor in series with a low resistance copper wire coil to obtain the desired value of resistance in the coil circuit.

Very fine copper wire with many turns is often used so that the coil will have the required circuit resistance. A rarely used, but more expensive, alternative is to use high DC resistance wire such as nichrome (nickel/chrome) wire for winding a coil.

An example of a DC voltage being applied to a coil is in the use of a DC operated electromechanical relay. The DC resistance of the relay coil circuit acts to provide sufficiently high resistance to limit the coil current to an appropriately low value. Either a resistor in series with a low resistance coil or a coil wound with many turns of fine copper wire is used.

The resulting coil current will still produce a magnetic field strong enough to actuate the relay when a DC voltage is applied across the terminals of the coil.

FIGURE OF MERIT OR QUALITY FACTOR (Q)

For inductors used in tuned circuits, the DC resistance of the wire provides another characteristic called its *figure of merit*, or *quality factor* (Q). The Q of a coil is the relationship between its inductive reactance (**X_L**) and its DC resistance (**R**) and is expressed mathematically as the ratio between these two separate values:

$$\text{Figure of Merit} = \text{Quality Factor} = Q = \frac{X_L}{R}$$

- A coil with relatively low DC resistance has a high Q
- A coil with relatively high DC resistance has a low Q

The Q of a coil establishes the response of a tuned or resonant circuit (a coil connected to a capacitor). A high Q coil in a tuned circuit provides a very selective (sharp) response to a single frequency, rejecting all other frequencies except the one selected. A low Q coil in a tuned circuit provides a wide response to the desired frequency, including the lower and higher frequencies adjacent to it. The extent of the response depends on the Q of the coil. (See Figure 12.3)

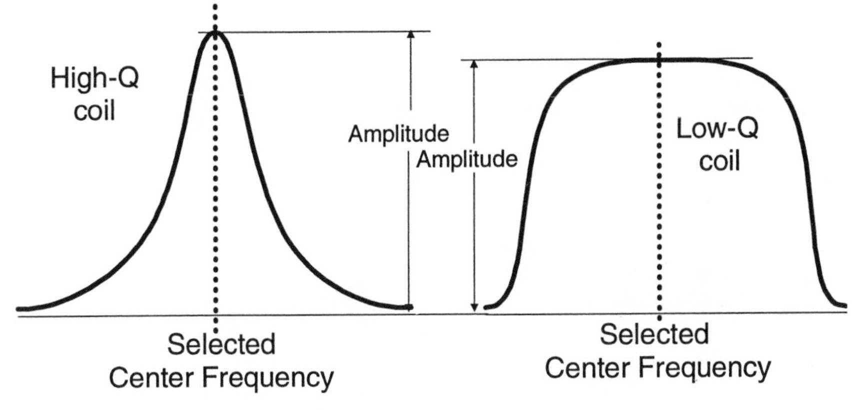

Figure of Merit - Q
Figure 12.3

INDUCTOR SPECIFICATIONS

INDUCTANCE (L) - Specified as its nominal value at +25°C measured in henrys (H), millihenrys (mH), or microhenrys (μH)

TOLERANCE - Specified as a percentage of the value at +25°C

MAXIMUM OPERATING VOLTAGE AND CURRENT - Specified in volts and amperes under continuous operation

POWER RATING - Specified in watts at +25°C with derating information included

DC RESISTANCE - Specified in ohms at +25°C

DIELECTRIC STRENGTH OR DIELECTRIC WITHSTANDING VOLTAGE (DWV) - Specified as the maximum voltage applied between either terminal and the frame at +25°C and at sea level

CORE MATERIAL - Selected for the frequency range required

OPERATING AND STORAGE TEMPERATURE RANGE
- For military and space applications: -55°C to +125°C
- For commercial, industrial, and consumer applications: No specific temperature range - typically, 0°C to +85°C

TEMPERATURE COEFFICIENT - Specifies the percent change in the value of inductance at +25°C per degree C change in temperature

PACKAGE INFORMATION - Includes the overall package size, weight, mounting features, type of enclosure, nature of the shielding against RFI, lead material, and terminal type

ENVIRONMENTAL CAPABILITIES - Specifies the inductor's resistance to environmental stress, such as: humidity, salt spray, fungus, chemical solvents, shock, vibration, etc.

INDUCTOR APPLICATIONS

Since an inductor is a frequency-dependent component, it has different values of inductive reactance at different frequencies. This characteristic provides the inductor with a variety of applications when used with pulsating DC and AC voltages.

Depending on the specific inductor that is used, different kinds of filters, tuned circuits, and other circuits can be created. They consist of single or multiple inductors used either alone or with one or several capacitors.

DC FILTERING - A power choke in conjunction with filter capacitors to efficiently change pulsating DC to steady-state DC.

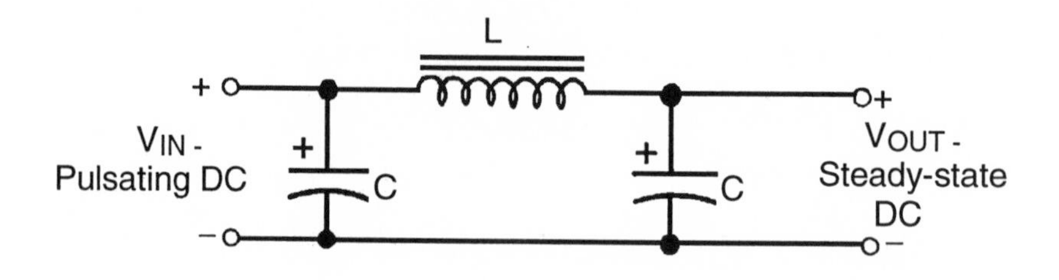

DC Filtering - Changing Pulsating DC To Steady State DC
Figure 12.4

The circuit shown in Figure 12.4, illustrates a low DC resistance inductor (choke) in conjunction with two polarized capacitors functioning as a pulsating DC filter. With many turns of wire wound around an iron core, the choke has a high inductive value and, therefore, a high inductive reactance even at the 60 Hz power line. Using a choke instead of a resistor between the two capacitors offers more efficient filtering, but is more costly.

The high inductive reactance of the choke acts to increase the discharge time of the first capacitor providing a considerable reduction of ripple (voltage variation) present in the initial steady-state D.C. voltage. The choke's neglible DC resistance minimizes the voltage drop and I^2R loss would be produced if an equally high resistor were connected between the two capacitors.

AC FILTERS - An AC filter is an inductor/capacitor circuit, or a multi-inductor circuit which accepts or rejects specific frequencies.

- A low pass filter allows only low frequency AC voltages to pass through a circuit. (See Figure 12.5)

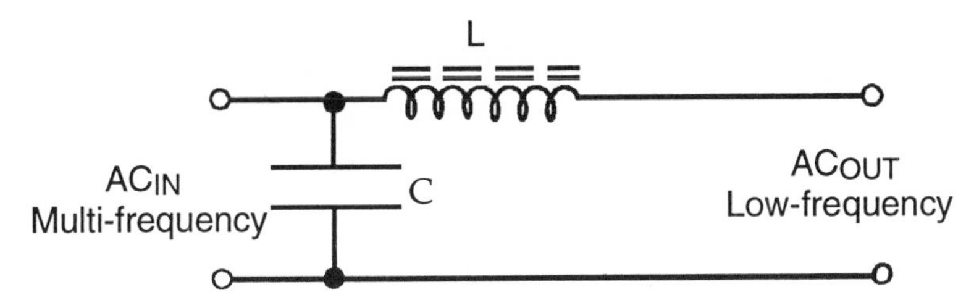

Low Pass Filter
Figure 12.5

- A high pass filter allows high frequency AC voltages to pass through a circuit. (See Figure 12.6)

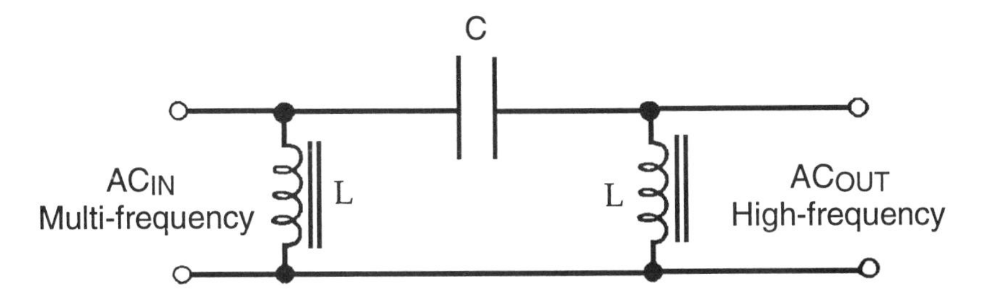

High Pass Filter
Figure 12.6

- A band pass filter allows AC voltages within a specific band of frequencies to pass through a circuit. (See Figure 12.7)

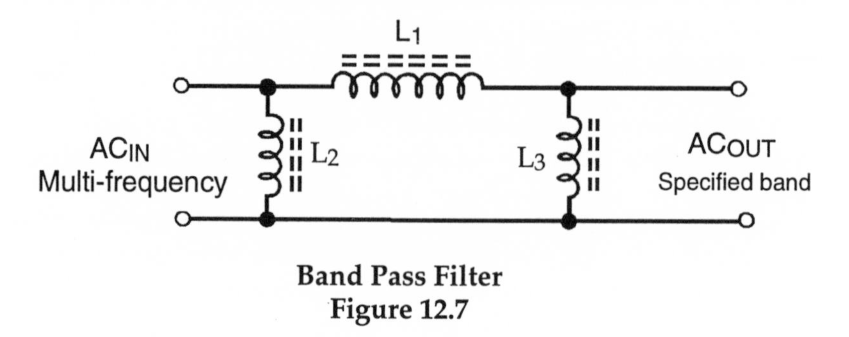

Band Pass Filter
Figure 12.7

TUNED (RESONANT) CIRCUITS - When used in conjunction with a capacitor to select (tune to) a specific frequency, these circuits attenuate frequencies above and below a desired frequency.

- A *series resonant circuit* consists of an inductor in series with a capacitor. The values of L and C are chosen for a specific frequency at which the inductive reactance (X_L) is equal to the capacitive reactance (X_C). In this condition, the two reactances oppose each other and cancel each other.

  At this point, the circuit is then in *resonance;* the equivalent reactance (AC resistance) of the circuit is zero and the AC resistance of all other frequencies is extremely high. Only the signal having the desired frequency will be at the output of the circuit. (See Figure 12.8a)

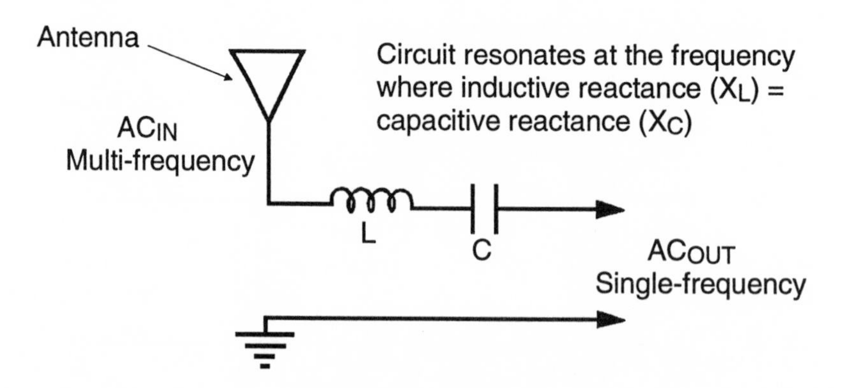

Series Resonant Circuit
Figure 12.8a

- A *parallel resonant circuit* consists of an inductor and capacitor connected in parallel. (See Figure 12.8b) The component values are selected so that, at a specified frequency, the two reactances are equal ($X_L = X_C$). At this point, the circuit is in resonance.

  In resonance, the reactances aid each other with the equivalent reactance approaching infinity. With no loading effect at the desired frequency, the selected signal is transferred to succeeding stages. The circuit presents a very low equivalent reactance to all other signals with frequencies other than the desired one, effectively shorting the undesired signals to ground.

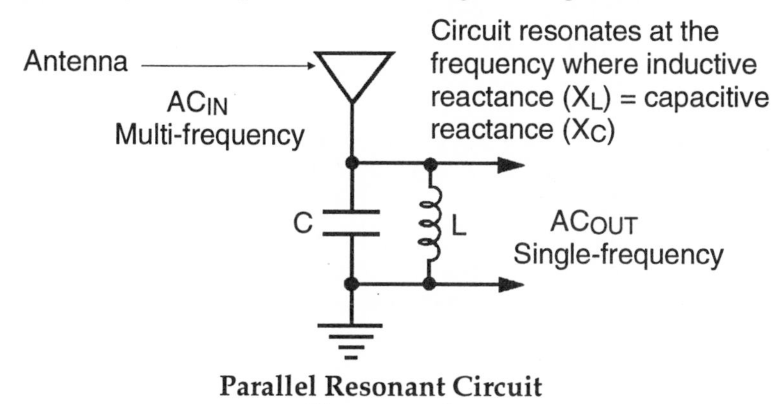

Parallel Resonant Circuit
Figure 12.8b

A practical resonant circuit consists of either a variable inductor or a variable capacitor to allow selection of a specific frequency over a wide range of input signals. A variable capacitor is more commonly used as the tuning component.

Depending on the Q of the inductor, the resonant circuit could be sharply tuned to a single, selected frequency or broadly tuned to a group of frequencies around a center frequency. (See Figure 12.3)

RF CHOKE - Since an inductor provides very high AC resistance to high frequencies ($X_L = 2\pi fL$), the inductor used at radio frequencies (RF) is referred to as an *RF choke* and is made with a ceramic or air core. The function of an RF choke is to resist the flow of RF currents.

In the circuit of Figure 12.9, the RF choke presents a very high AC resistance to RF, keeping the high frequencies in the amplifier from being inserted into the DC power supply. Since the supply may be common to other sections of a system, it is important that it be kept free of any RF that can interfere with other circuits in the system.

To further enhance the capability of the RF choke circuit, an optional bypass capacitor (C) can be used across the DC voltage supply.

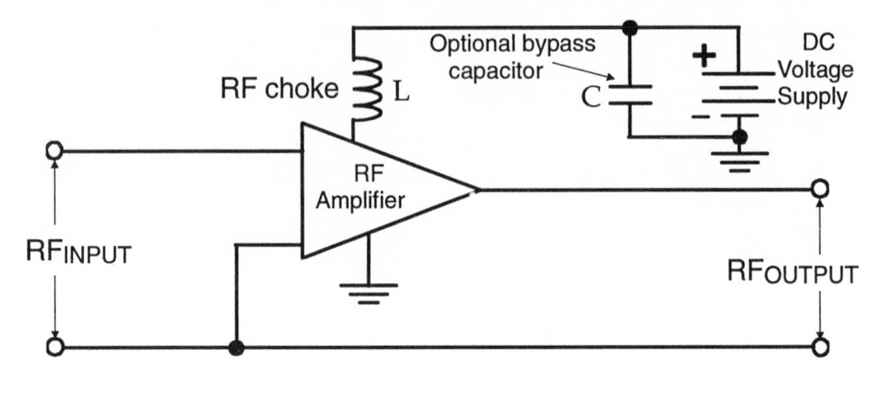

RF Choke Circuit
Figure 12.9

PASSIVE DELAY LINE - a group of interconnected inductors and capacitors specifically configured to delay a linear or digital signal on a transmission line or circuit for a predetermined time.

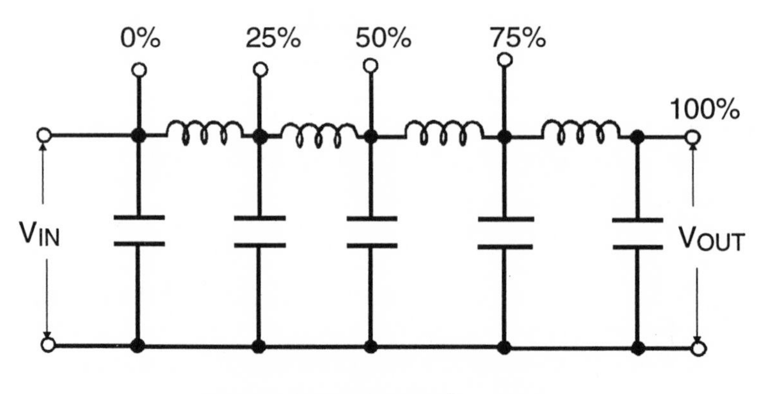

Tapped Passive Delay Line
Figure 12.10

Passive delay lines are inductor-capacitor (LC) networks designed with different LC values, core materials, and packages. (See Figure 12.10) They provide a wide variety of time-delay possibilities (ranging from 1 nanosecond to 60 milliseconds) and attenuation factors between input and output. These devices are generally supplied as multi-tapped delay lines to provide the options of 10% to 100% delay with a specified attenuation limit. Typically, they are assembled in a 14 pin dual in-line package (DIP).

IMPEDANCE (Z)

The true resistance of a circuit or load is a complex parameter called **impedance**. It is measured in ohms and designated with the letter "Z". Impedance is the combination of capacitive reactance (**X_C**), inductive reactance (**X_L**), and the DC resistance (**R**).

$$\textbf{Impedance (Z)} = \sqrt{R^2 + (X_L^2 - X_C^2)}$$

In a purely capacitive circuit, the charging capacitor current **leads** the capacitor voltage by 90°. (See Figure 12.11a) As DC resistance is added, the phase difference between current and voltage decreases until X_C is negligible compared to the DC resistance. At this point, capacitor current (I_C) and capacitor voltage (V_C) are in phase.

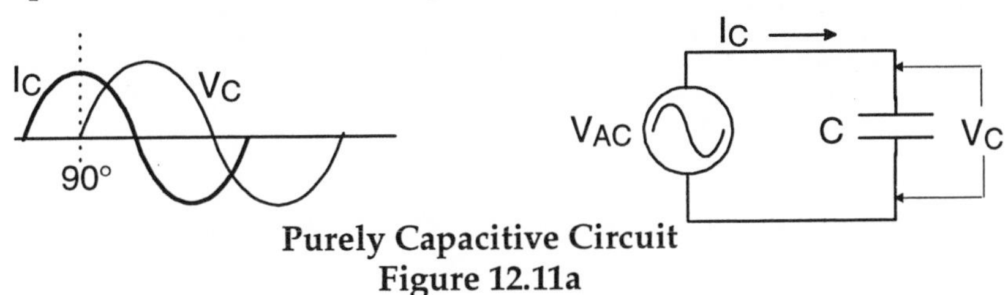

Purely Capacitive Circuit
Figure 12.11a

In a purely inductive circuit, the current in the inductor **lags** the voltage across the inductor by 90°. (See Figure 12.11a) As resistance is added, the phase difference between current and voltage decreases until X_L is negligible compared to the DC resistance. At this point, inductor current (I_L) and inductor voltage (V_C) are in phase.

Purely Inductive Circuit
Figure 12.11b

In the AC circuit of Figure 12.11c, **Z** is calculated as shown. Since X_L and X_C are 180° out of phase, they cancel each other. At a specified frequency, **$X_L = X_C$**, the circuit is in resonance, and **Z = R**.

$$Z = \sqrt{R^2 + (X_L^2 - X_C^2)}$$

AC Circuit
Figure 12.11c

When the impedances of two connected circuits or loads are matched, a maximum transfer of power between the two is achieved. A mismatch of impedance characteristics will produce little or no power transfer. Some examples of required impedance matching are:

- Between a computer input or output and a telephone line
- Between the output of an amplifier and a loudspeaker
- Between two circuits, each having a different Z_{OUT} and Z_{IN}

TRANSFORMERS

A **transformer** is a component consisting of two or more coils of wire wound around a common core to efficiently transfer electrical power or signals between circuits while providing electrical isolation between these circuits. This transfer of power is accomplished through the process of *magnetic induction.*

TRANSFORMER FUNCTIONS

- A *power* transformer transfers power (voltage times current) from an AC power line to the section of an electronic system that changes AC to pulsating DC (rectifier section). It isolates the lethal power line from the electronic circuit and acts to change the level of the power line voltage to a desired amplitude.

- An *output* transformer transfers the signal from the output of an audio amplifier to a loudspeaker while matching its electrical characteristics.

- A *radio frequency* (RF) transformer is an efficient coupling device that transfers high frequency sine waves between stages of a radio frequency amplifier circuit while providing high voltage isolation between the windings.

- A *pulse* transformer efficiently couples digital pulses between stages of a digital system while providing isolation between these stages.

- In most applications, the isolation feature of a transformer separates circuits that are sensitive to common electrical noises or voltage transients. In addition, because of its isolation characteristic, a transformer provides protection from exposure to direct contact with a lethal power line.

TRANSFORMER CONSTRUCTION AND OPERATION MAGNETIC INDUCTION

The operation of a transformer is based on producing a changing magnetic field in one of its sections to allow the power in that magnetic field to be efficiently transferred to a second section.

- Construction of a transformer begins with a length of insulated copper wire wound around a core. This winding is called the *primary winding*, or *primary*. The exact number of turns in the primary must be specified.

- Plastic or other insulating material is used to encapsulate the primary, and the ends of the winding are connected externally to terminals. A second length of insulated copper wire is now wound around the encapsulated primary. Here, too, the exact number of turns is specified. This is called the *secondary winding*, or *secondary*. The assembly is then encapsulated with insulation and the secondary ends are connected externally to terminals.

- If required, additional secondaries can be wound around this assembly, with each secondary having a different number of turns than the first secondary or, if indicated, the same number of turns. No electrical connection exists between the primary and secondary windings since each of the secondaries has insulating material between it and the primary winding.

- The transformer is then enclosed in an open frame case, or in other appropriate cases, with all terminals made externally accessible. (See Figure 12.12)

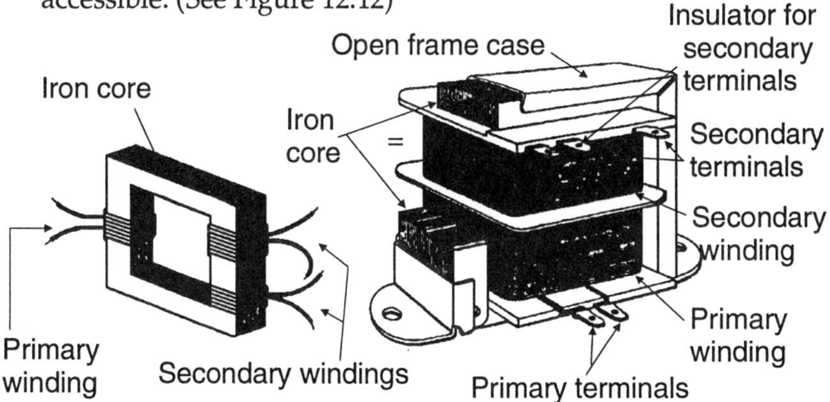

Typical Transformer Construction
Figure 12.12

The relationship between windings is called the *turns ratio* and is expressed as a ratio between the number of turns in the primary winding and the number of turns in the secondary winding. This ratio can be either:

- A *step-up ratio* indicating that there are more turns in the secondary than in the primary
- A *step-down ratio* indicating that there are fewer turns in the secondary than in the primary
- A *one-to-one ratio* indicating that both primary and secondary have the same number of turns

When AC voltage is applied to the primary, causing alternating current to flow, the current through the primary will create a changing magnetic field around the primary winding.

The changing magnetic field, or the changing lines of magnetic force, will cut across the secondary winding and *induce* an alternating voltage and current in the secondary. This phenomenon is called *magnetic induction.*

The voltage induced in the secondary will depend on two factors:

- The voltage applied to the primary
- The primary-to-secondary turns ratio: N_P/N_S

Each turn of the primary winding produces a specific portion of the magnetic field. Conversely, a given magnetic field, through magnetic induction, creates a specific value of voltage in each turn of the secondary winding. As result, the following relationships exist:

$$\frac{\text{Primary voltage } (E_P)}{\text{Primary turns } (N_P)} = \frac{\text{Secondary voltage } (E_S)}{\text{Secondary turns } (N_S)}$$

then, $\mathbf{E_P/E_S}$ (the voltage ratio) = $\mathbf{N_P/N_S}$ (the turns ratio)

For purposes of simplification, it can be assumed that there are no losses in the transformer, since a transformer has efficiencies of 96% to 99%. Therefore, the power in the primary (P_P) is equal to the power in the secondary (P_S).

Since: $P = E \times I$, then $\mathbf{E_P \times I_P = E_S \times I_S}$

By transposing terms, the following relationships are produced:

$$E_P/E_S = I_S/I_P = N_P/N_S$$

- The ratio between the primary and secondary voltage is **directly** proportional to the turns ratio.
- The ratio between primary and secondary current is **inversely** proportional to the turns ratio.

When an AC voltage is applied to the primary terminals, the secondary voltage, compared to the primary voltage, will either increase or decrease according to the turns ratio. The secondary current compared to the primary current will change in an inverse manner by the same ratio.

Figure 12.14 illustrates a circuit having an AC voltage applied to the terminals of the primary and a load connected to the secondary terminals. This voltage can be in the form of AC (or full-wave pulsating DC), but **not** steady-state DC. Steady-state DC voltage maintains a constant level of DC, and the resulting magnetic field will also be a constant value.

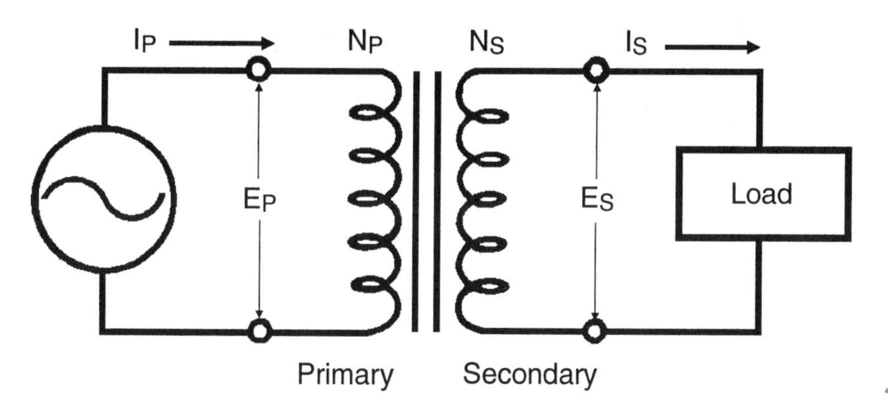

Secondary Voltage and Current vs. Turns Ratio
Figure 12.13

Power cannot be transferred with a constant DC voltage applied to the primary winding since the magnetic field created by steady-state DC is constant. The transformer must have a constantly varying voltage applied to the primary to function properly.

TRANSFORMER CORE MATERIAL

A transformer is a frequency-dependent component that is similar to two or more inductors wound on the same core; the selection of its core material depends on the frequency at which it operates. At lower frequencies, more ferrous material is required, while at radio frequencies, a nonferrous (plastic or ceramic) or air core is used.

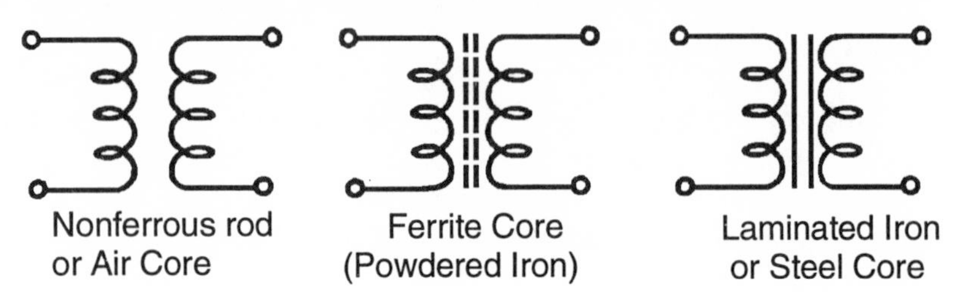

Transformer Symbols with Different Cores
Figure 12.14

- At frequencies up to 20 kilohertz, the core material is made of iron or steel laminations.
- At the mid-frequencies, from 20 kilohertz to about 10 megahertz, the transformer core is made of ferrite or powdered iron.
- At frequencies above 10 megahertz, the core could be a nonferrous rod or cylinder or an air core (a free-standing transformer without a physical core).

SECONDARY TAPS

As the secondary is being wound, connections (taps) may be made from any point on the secondary to external terminals. These connections are made by scraping away the insulation on the secondary winding at the appropriate points and connecting conductors from these points to corresponding terminals.

A tap placed exactly half way down the secondary winding is called a *center-tap* (CT), however, there can be taps set at the 10% point, 90% point or anywhere along the secondary winding. A transformer must have a primary and, at least, one secondary. It also may have multiple secondaries, each one with a specific turns ratio with respect to the primary winding. Each secondary may have a single tap, multiple taps, a center tap, or no tap.

MULTIPLE-TAPPED SECONDARY

With multiple taps on the secondary, a range of voltages can be selected from the total secondary voltage, depending on the number of taps available.

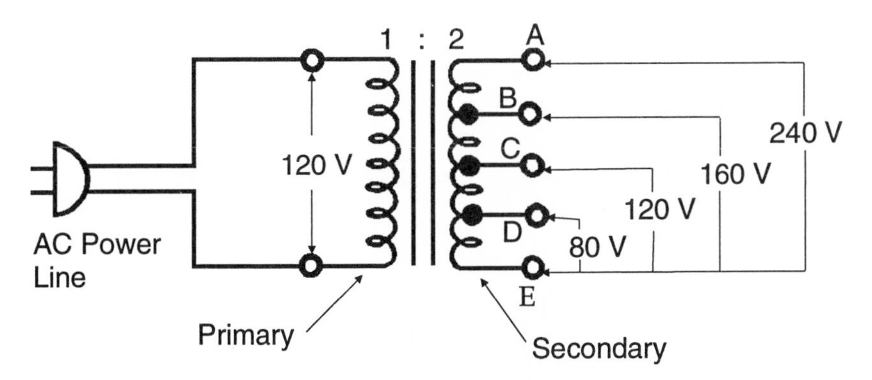

Secondary with Multiple Taps
Figure 12.15

The step-up transformer in Figure 12.15 has three taps on the secondary in addition to its end terminals. If the total secondary voltage from end to end (point A to point E) is 240 volts, the voltages available at the other secondary terminals are:

- 160 volts (point B to point E)
- 120 volts (point C to point E)
- 80 volts (point D to point E)

The 120 volt terminal (point C) is also the center-tap; it is located exactly half-way between points A and E.

A transformer with a multi-tapped secondary is designed to:

- Provide several levels of secondary voltage since different parts of a system may need different AC and DC voltage levels.
- Offer a more universal power transformer for a variety of different secondary voltages.
- One transformer can function in a variety of different applications if several secondary voltages are required.

CENTER-TAPPED SECONDARY

A center-tapped secondary provides a means of making available all or half the total secondary voltage. It can connect either of two different loads to the output of a circuit when the required voltage of one load is exactly one-half the other.

- In the circuit of Figure 12.16a, a load is connected across terminals A and C of the secondary with the switch set to point A.
- When 120 volts AC is applied across the primary and with a one-to-one turns ratio of the transformer, the voltage across $Load_1$ is also 120 volts AC. Half that voltage is required for $Load_2$.

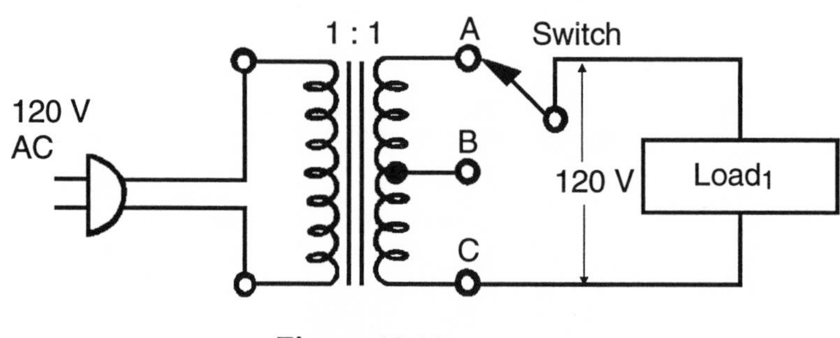

Figure 12.16a

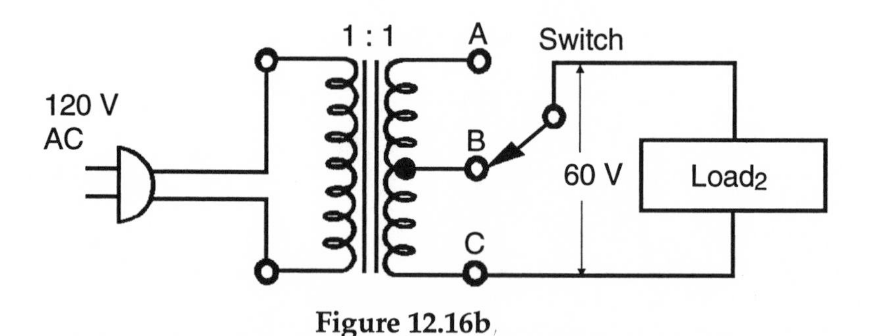

Figure 12.16b

Using The Center-tap to Provide Half the Secondary Voltage

In the circuit of Figure 12.16b, the same transformer has the same AC voltage applied to the primary. If the switch is set to its alternate position (point B on the secondary), with $Load_2$ connected between the center-tap and bottom of the secondary (across points B and C), the required 60 volts AC is applied across the load.

An additional use for a center-tapped secondary is where the center-tap terminal is used as a reference (ground). At the outer terminals of the secondary, two equal AC voltages are created having opposite momentary polarities with respect to ground.

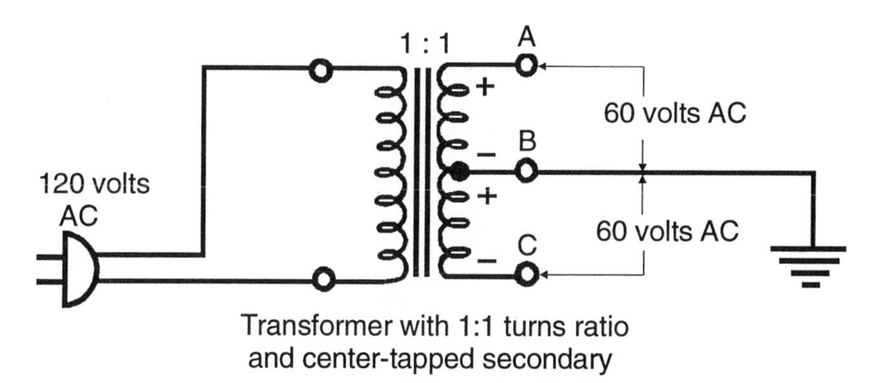

Secondary Center-Tap Used as a Reference Point
Figure 12.18

When an AC voltage is applied across the primary, two AC voltages are induced across the secondary - the voltage from point A to point B and the voltage from point C to point B.

At any instant, a center-tapped secondary provides two separate and opposite polarity voltages with respect to a common reference. This feature allows for full-wave rectification in a DC power supply application. Full-wave rectification is discussed in detail in Volume Two - Part I - Discrete Semiconductors.

TRANSFORMER SPECIFICATIONS

FREQUENCY OF OPERATION - Specified at a nominal frequency or over a range of frequencies

NOMINAL PRIMARY VOLTAGE AND CURRENT - Specified for continuous use at +25°C

SECONDARY VOLTAGE AND CURRENT - If two or more taps exist at the secondary (or several secondaries), their locations on the secondary are specified.

TURNS RATIO - Specified as the number of turns in the primary winding compared with the number of turns in the secondary winding(s)

PRIMARY AND SECONDARY DC RESISTANCE - The DC resistance of the wire in the primary and secondary windings specified in ohms

DIELECTRIC STRENGTH - Indicates the quality of the insulation between the windings and between the windings and the core. It is specified as the maximum voltage the insulating material can withstand without breakdown. This specification is also referred to as INSULATION STRENGTH.

OPERATING AND STORAGE TEMPERATURE RANGE - For military and space equipment, from -55°C to +125°C; specified by the appropriate military specification. For commercial, industrial, and consumer use, the range is typically from 0°C to +80°C.

MAXIMUM INTERNAL TEMPERATURE RISE - Specified in °C at rated operating conditions

PHYSICAL CHARACTERISTICS - Includes its size, weight, type of case, terminal spacing, dimensions, and mounting features. These specifications are generally supplied in the manufacturer's data sheet or the customer's specification control drawing.

COMMON TRANSFORMER TYPES

POWER - Operates at 50 to 400 hertz at a nominal line voltage from 105 to 130 volts. It is manufactured with single or multiple secondaries and is available in a wide range of step-up and step-down turns ratios. The secondary(s) can have a single tap, multiple taps, or no tap and some units are made with a tapped primary to accommodate a range of source voltages.

Output voltages can range from 3 volts to several thousand volts with output currents from 0.01 to 1500 amperes. Cores are iron or steel laminations. The transformer can be enclosed in a hermetically sealed package for military or space applications or in an open frame or plastic enclosure for commercial, industrial, or consumer use.

PULSE (RF) - Couples square wave pulses (with emphasis on fast rise and fall times of the pulse) and high frequency response. This transformer is packaged in a miniature enclosure and uses a nonferrous core. (See Figure 12.19)

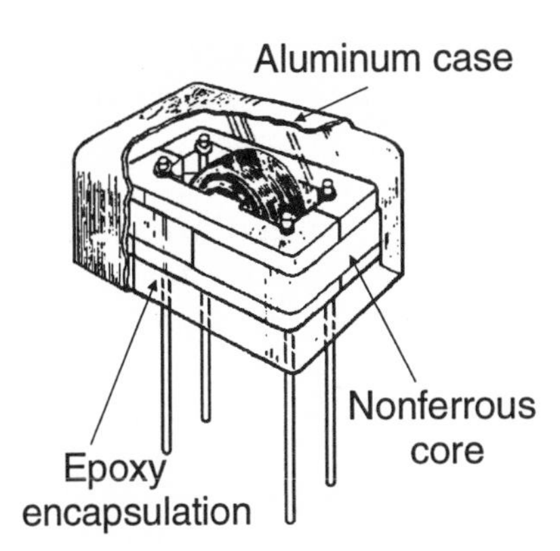

Pulse Transformer
Figure 12.19

ISOLATION (POWER) - A power transformer with a one-to-one turns ratio between primary and secondary which provides isolation between the line voltage and the secondary load. Usually, an isolation transformer includes a *Faraday shield,* a screen of non-magnetic metal wound between the primary and secondary windings and connected to the transformer core. The shield prevents capacitive coupling of spurious signals and minimizes noise between windings. A Faraday shield, however, may increase leakage current resulting in reduced transformer efficiency.

FERRO-RESONANT OR CONSTANT VOLTAGE (POWER) - This type operates from a varying AC power line (from 105 to 130 volts) to provide a constant AC output voltage. Typical output voltages may be specified from 6 to 118 volts AC, regulated to within ± 1%. Line voltage frequency variations of more than 1 hertz above or below the nominal frequency cannot be tolerated without having the transformer go out of regulation.

CONTROL (POWER) - A small power transformer used for powering a control device such as a relay, solenoid, or other low voltage AC control component. Typical output voltages are 12 and 24 volts AC at current capabilities of 4 to 16 amperes.

AUDIO - This transformer is used to match the output impedance of a power amplifier to the impedance of a loudspeaker to provide maximum transfer of power and uniform response over the desired frequency range. In high fidelity audio systems, audio transformers operate from 20 Hz to 20 kHz. In audio systems involved with voice communications only, they operate from 200 to 3000 hertz.

RADIO FREQUENCY - A transformer designed for operation at a fixed high frequency in conjunction with a capacitor across either the primary, secondary, or both, to create a tuned or resonant circuit. Most types use air cores, however, some are made with a ferrite movable core to permit adjustment of inductance over a given range. They are usually assembled in an aluminum shielded can to reduce pickup or radiation of magnetic fields. Generally, the capacitors are mounted in the same can as the transformer and are connected internally to the primary and secondary windings.

SYNCHROS AND RESOLVERS - Transformers that resemble small AC motors; used as transducers to measure, control, or transmit angular position, speed, or acceleration. These types have iron cores.

TESTING OF MAGNETIC COMPONENTS

Hi-pot (high-potential) testing is a quality control technique that establishes a standard of safety for consumers when using magnetic components. The purpose of these tests is to detect the presence of faults in the insulation system of a magnetic component and to eliminate the possibility of shock and fire hazard.

- At least ten times the rated voltage is applied for one minute between the primary winding and secondary winding(s) and the metal frame of the device. The purpose of this test is to establish a sense of confidence in the product by safely applying a voltage that is considerably greater than one to which the product will ever be normally subjected. The concept supporting this test is that if the product can withstand the excessive voltage for one minute, it can be expected to handle the line voltage and transient overloads throughout its natural life.

- Hi-pot testing verifies insulation quality and adds to the reliability of the product by detecting workmanship errors. It can also detect if the wire insulation is being pinched, stressed, or ruptured, each of which can be a cause for rejection.

- Hi-pot testing appears in almost all specification drawings for a magnetic component. Passing the hi-pot test generally will assure Underwriters Laboratory (UL) approval for magnetic component safety standards.

REINFORCEMENT EXERCISE

Answer TRUE or FALSE

1. Inductor is the generic term for a component consisting of a wire wound around a core producing an inductance in the device. Other terms for this component are coil, or choke, and more clearly defines an inductor's application.

2. The amount of inductance in an inductor, coil, or choke is a function of the number of turns and the core material. The greater the number of turns or the more iron in its core, the higher the inductance.

3. Inductance produces resistance to AC, called inductive reactance (X_L). X_L depends on the frequency (f), the amount of inductance (L), and the mathematical constant of 2π (6.28).

4. Inductive reactance decreases as frequency increases.

5. Inductors, coils, and chokes are made with cores of iron or steel laminations, powdered iron, ferrite cores, or air cores, with its selection depending on the frequency of operation.

6. Ferrite core inductors are extremely efficient at frequencies below 100 hertz.

7. Filtering of pulsating DC obtained from the 60 hertz power line is accomplished most effectively with a choke made with a core of iron or steel laminations.

8. RF chokes are not very effective in isolating high frequency signals from the supply voltage, but are used as part of a tuned circuit.

9. The voltage at the output, or secondary, of a transformer varies inversely with the turns ratio of the transformer. The current in the secondary varyies proportionately with the turns ratio.

10. One significant function of a transformer is its ability to change levels of voltage and current between the primary and secondary while providing electrical isolation between the circuits to which the transformer is connected.

11. Transformers are used to match the impedance of two circuits to provide the most efficient transfer of power from one to the other.

12. Transformers can couple AC voltages, pulsating DC voltages, digital pulses, and steady-state DC voltage.

13. A transformer can have a single secondary, or many secondaries, with or without single taps, or multiple taps on the secondaries.

14. The dielectric strength of a transformer is specified as the maximum voltage its insulation can withstand without rupturing. These voltages are specified between the primary and secondary windings and between the windings and the transformer frame.

15. Hi-pot testing is a quality control technique that establishes a standard of safety in magnetic components.

16. To verify the quality of the insulation, the AC voltage that is applied as a hi-pot test is no greater than the rated voltage of the transformer secondary and lasts for no longer than one second.

17. In addition to verifying the insulation quality, hi-pot testing detects workmanship errors and is used to determine if the insulation is pinched, stressed, or ruptured.

Answers to this reinforcement exercise are on page 281.

13 CHAPTER THIRTEEN

MISCELLANEOUS PASSIVE COMPONENTS

MISCELLANEOUS PASSIVE COMPONENTS

CONNECTORS

A **connector** is a mechanical coupling device that provides rapid joining or separation between two sections of an electrical or electronic circuit. A connector pair consists of two mating parts - a *plug* and a *receptacle*. The plug and receptacle must match but either may be a male or female configuration.

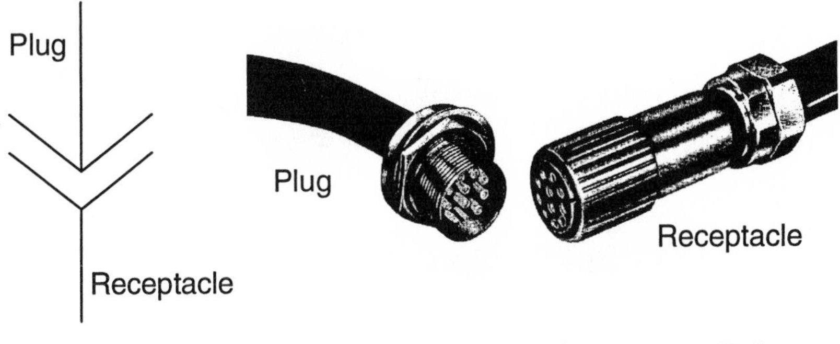

Connector Symbol **Typical Connector Pair**

Figure 13.1

Each plug and receptacle generally contains:

- A set of contacts that provide electrical continuity between the two mating sections
- A set of terminals that electrically connect a cable or conductor to the contacts
- Insulating material to electrically isolate the contacts and terminals
- A housing in which the contacts, terminals, and insulation are assembled
- A mechanical arrangement to join the two sections
- Protection against mechanical and environmental stress

When the plug and receptacle are mechanically mated, the electrical connection is established between the two parts of the circuit. Conductors are connected to the terminals of the two connector sections by soldering, crimping, wire-wrapping, or with screw terminals.

The receptacle section is normally mounted in a permanent position in the equipment while the plug section is free to move when disconnected. *In-line connectors,* however, join one cable to another while both sections are free to move.

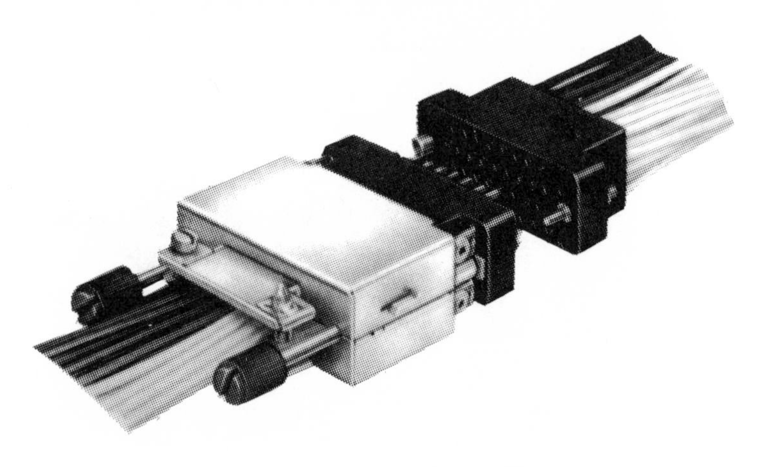

In-Line Connector
Figure 13.2

STANDARDIZATION AND INTERCHANGEABILITY

Electronic systems contain a wide variety of connectors and interconnecting devices. There are hundreds of basic types of connectors; the variations are endless and can be quite complex.

Most passive components have been standardized with regard to their electrical and mechanical parameters and, generally, there is a high level of interchangeability among all passive components, including connectors. A problem in connector standardization is that these components are often involved in interfacing with a wide diversity of circuit configurations and require a concomitant variety of connectors to satisfy these configurations. Total connector component standardization and interchangeability is a goal that may be quite difficult, if not impossible, to attain.

CONNECTOR CATEGORIES

Connectors can be categorized in several ways or in combinations of these different ways. Two major categories include **function** and **operating frequency**.

FUNCTION:

- Internally (inside an enclosure) - Interfacing with:
  2 in-line cables; 2 PC boards; 1 PC board and 1 internal cable

- Externally(outside an enclosure) -Interfacing with:
  1 cable and 1 fixed receptacle

OPERATING FREQUENCY:

Connectors are required for a wide variety of control, data, and communications signals operating with DC voltages and AC voltages at all frequencies. At lower frequencies, the requirements are less critical for both connectors and interconnecting cables.

Depending on the operating frequency, a specific type of connector optimizes the transfer of information and/or power by connecting between circuits. Both the connector and interconnecting cable must be part of the overall connector design.

- At power line frequencies (50 to 400 Hz), a power connector is selected on the basis of the amount of heat produced in the connector. This heat is created by the flow of current through the contact resistance; the operating temperature should not exceed its maximum rating. To minimize heat-related damage, power connectors use large contact pins, large connecting areas, firm contact pressure, and low-resistance interface materials.

- Phono plugs and jacks are selected for two-conductor or twisted pair cables in the audio range (20 Hz to 20 kHz) and for VCR applications up to 6 MHz. Coaxial cable connectors are utilized in the ultrasonic range (20kHz to 1 MHz), especially as the frequency approaches 1 megahertz, and for RF up to 600 MHz.

- At radio frequencies (1 MHz and above), unique connector considerations exist. High frequency capacitance, created by the spacing between the contacts and between the terminals of the connector, and the effects of the connector and/or cable inductance, must be minimized. At these frequencies, coaxial connectors (also called RF connectors) are used to maintain appropriate matching characteristics and minimal signal loss.

- A microwave connector is used at microwave frequencies (1 GHz and above) and is not just a device joining two parts of a circuit but is also an integral part of that circuit. Its shape, size, composition, and configuration all contribute to circuit operation.

- Special fiber optic connectors in the visible light range (700 to 350 nanometers) minimize coupling losses.

- Edge-card connectors handle low voltages (3.3 to 24 volts) and low currents (1 to 20 milliamperes) in digital systems. Digital pulses operate from DC to frequencies up to 100 MHz.

- Carefully matched RF connectors with extremely low capacitance are required in complex systems operating at 1 GHz and above to maintain the integrity of these signals for timing and sequencing.

CONNECTOR SPECIFICATIONS

ELECTRICAL CONSIDERATIONS

CURRENT - Specifications include maximum surge current and continuous operating current for the connector contacts. The current requirements of the conductors attached to the connector terminals determine conductor size and material.

VOLTAGE - Specifies the maximum allowable surge voltage and maximum allowable working voltage between the contacts and between the terminals. The spacing between the terminals and between the contacts determines the voltage values.

DIELECTRIC WITHSTANDING VOLTAGE RATING - Specifies the maximum allowable voltage across the insulating material between the terminals and the shell, frame, and body of the connector.

RADIO FREQUENCY SHIELDING - If eliminating or minimizing radio frequency interference (RFI) is necessary, terminating or grounding the shielding material of the cable and the connectors must be considered.

OPERATING FREQUENCY - Usually dictates the specific kind of connector required. Different connector types are available that cover the entire frequency spectrum, from DC, to audio, to RF, to microwave, and to the light frequencies. Depending on frequency requirements, these connectors can call for special construction, assembly, size, and mounting features.

MECHANICAL CONSIDERATIONS

TERMINALS - Specifications include number, size, center-to-center spacing between terminals, material, style, and recommended wire size (stranded or solid), and configuration. Where required, some means of maintaining proper polarity of the terminals is given.

Terminal styles include dip-solder types, solder eyelets, wire-wrapping pins, crimping pins, and bifurcated tabs used in the *insulation displacement connector* (IDC) technique. Terminals are manufactured in a wide variety of sizes and shapes.

See Figure 13.3a for examples of several termination types. The cables for the terminals include jacketed and shielded jacketed types.

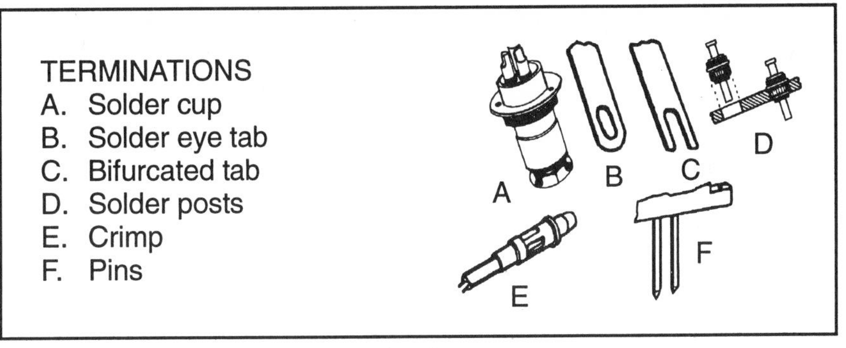

Terminal Types
Figure 13.3a

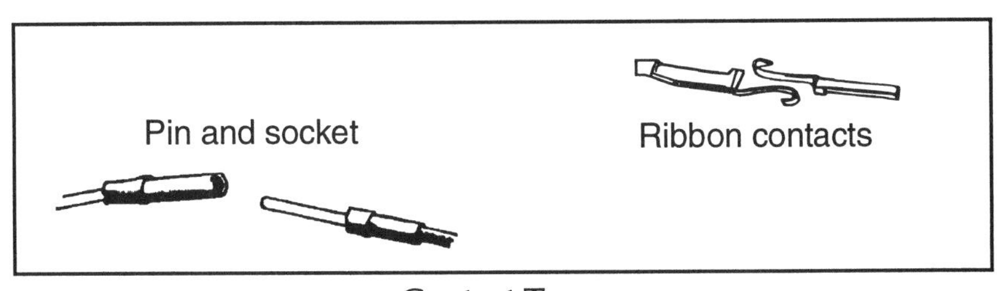

Contact Types
Figure 13.3b

CONTACTS - Selection of material depends on the application.

- **Beryllium copper** offers excellent mechanical properties, stable electrical and thermal characteristics, and high resistance to wear and metal fatigue. It is used for complex spring contact design.

- **Phosphor bronze** offers electrical features and mechanical properties required for simpler and effective designs at low cost.

- **Spring brass** is used where there are no requirements for high temperature operation or repeated flexing.

The surface areas at the point of contact between mating parts have an influence on the conducting ability of a connector. Dry circuit applications generally require gold-to-gold or special alloy surfaces. For contact surface-plating, thickness is specified. A variety of finishes are available, ranging from gold plating over nickel to gold alloys. Special geometries and techniques provide a gas-tight, high-pressure interface between the conductor and the connector contact.

Contact styles include bifurcated, cantilever, tuning fork, ribbon, and pin types. Style selection depends on the application. Examples of common contact types are shown in Figure 13.3b. Conventionally, an inserted pin or plug contact is called *male* and an accepting hollow pin or receptacle is referred to as *female*. Non-gender, or *hermaphroditic* contacts are those in which both mating members are exactly alike without differentiation by gender.

INSERTION AND EXTRACTION FORCE - These forces are needed to fully engage or to separate the mating connectors. The required forces depend on the number and size of contact pins, the strength of their springs, surface finish, and the rigidity of their mountings.

To a great extent, the required insertion force will be dictated by the life expectancy of the connectors, their intended use, the speed of insertion, and their immunity to shock and vibration.

- For minimum contact resistance, insertion force must be high, producing higher abrasion during insertion and removal. This results in reduced life of the contact surfaces.

- In test fixtures where connectors are engaged and disengaged frequently, *zero-insertion force* (ZIF) connectors are used to reduce wear on the contact surface and increase longevity.

MOUNTING FEATURES
These specifications include the connector's overall dimensions and the type of mounting, e.g.: full-mounting flange, half-mounting flange, or no mounting flange. In addition, strain relief provisions are included in cable connectors and other unique features are provided for panel mount, bulkhead mount, PC board mount, etc.

ENVIRONMENTAL CONSIDERATIONS

OPERATING TEMPERATURE

- For military and space equipment, -55°C to +125°C

- For commercial and industrial applications, typically, 0°C to +80°C

RESISTANCE TO ENVIRONMENTAL STRESS - The effects of corrosive materials, such as salt spray, fungus, and chemical solvents can be significant. If a connector is exposed to humidity, then the possibility of arcing, voltage breakdown, and leakage current exists.

- Provision must be made to protect the contact regions and conductor terminations against deteriorating effects. Protection may include face seals, rubber gaskets, "O" rings, and grommets. A potting compound might be required to seal the critical areas for equipment operating in severe environments.

- In some cases, the contact pins may require bonding to the connector by glass, or other materials, to provide an airtight (hermetic) seal against penetration by air, moisture, corrosive vapors, or other contaminants.

ALTITUDE AND WATER SUBMERSION FACTORS - Connectors are generally specified for operation at sea level. For operation at high altitudes or for operation when submerged in depths of water, certain electrical and mechanical specification trade-offs exist. This information is included on the manufacturers' data sheets and/or customers' specifications, when applicable.

Because of the multitude of connector types, sizes, terminals, contacts, and mounting variations, a clear and definitive set of specifications that will satisfy any required application should be determined after consultation with a connector manufacturer.

A partial list of the more popular connector types includes:

PC board card edge
Ribbon (Flat) cable
Solderless wire-wrapped
Insulation displacement (IDC)
Circular
Rectangular
Dual in-line (DIP) plug
Coaxial (RF)
Rack and panel
Zero-insertion force (ZIF)
D and D-subminiature
Microwave

Other types include: sockets, terminal blocks and strips, pins, plugs, jacks, binding posts, and variations of these devices.

Fiber optic connectors are a special group of connectors with specifications that are unique to fiber optic technology. They are discussed in Volume Two - Part Two - Optoelectronics.

GENERAL COMMENTS ON CONNECTORS

Connectors are often considered the least reliable link in the electronics chain and, even when carefully selected, properly assembled, installed, and shielded, they still can exhibit some unacceptable characteristics:

- At high frequencies, connectors can introduce resistance, capacitance, and inductance that could cause an impedance mismatch between the circuits they connect.

- Connectors provide points in a system where moisture, dust, and corrosive vapors could enter to disrupt circuit operation.

- If not properly fastened, connector sections can easily become disconnected, opening the circuits they should be connecting.

Ideally, when properly selected and installed in a system, connectors must be electrically invisible and identical to the characteristics of the conductors used to provide continuity. This condition is not always realized in actual practice.

The choice of connectors is most important in the overall design and operation of a system. It requires familiarity with system parameters and knowledge of the end-application. Selection and placement of a connector must be considered prior to the completion of a system and not installed into a system as a design afterthought.

INDICATOR LAMPS

Incandescent and neon lamps are passive indicators that visually signal a specific circuit condition in a system. An example is a POWER ON indicator lamp on the front panel of a system.

INCANDESCENT LAMP INDICATOR - Applying an AC or DC voltage across the filaments of an incandescent lamp causes current to flow. Current flow causes the filaments to heat and produce an emission of photons at a high temperature, radiating their familiar white light. This generation of light is called *incandescence.*

Incandescent light ranges over the light spectrum from non-visible infrared through the visible light range, emitting white light.

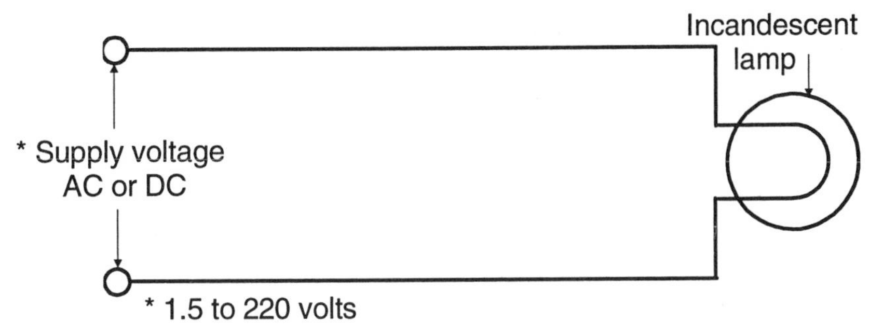

Incandescent Lamp Indicator Circuit
Figure 13.4

Incandescent lamps require from 1.5 to 220 volts, AC or DC, depending upon the specific type used. The corresponding current levels are such that the power consumed ranges from 1 to 7 watts per bulb. In itself, the typical 2 watts are considered to be low power, however, if one hundred incandescent lamps were assembled on a panel, a 200 watt power supply would be required.

A high power supply and the lamps themselves generate considerable heat that must be removed with heat sinks, fans, or other cooling devices. In addition, an incandescent lamp is not shock proof or vibration proof and has a longevity of about 2 years.

Because periodic replacement is required, soldering the lamp in place is not practical. A light socket is generally used, increasing parts count and production costs, while reducing reliability.

NEON LAMP INDICATOR - When fewer than 60 volts are applied across the terminals of a neon light indicator, the resistance of the neon gas is infinite, allowing no current to flow in the circuit. With no current flow, there is no light emission. An AC or DC voltage of at least 60 volts will cause the neon gas to *ionize* or break down.

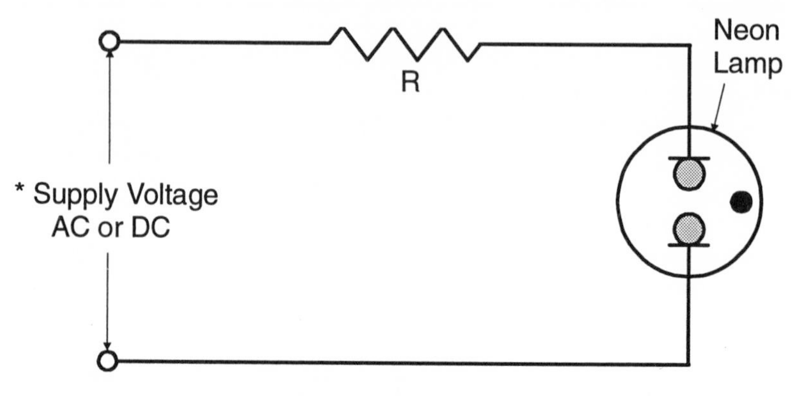

Neon Indicator Circuit
Figure 13.5

Ionization of the neon gas will change its initially infinite resistance to a low resistance value (essentially zero resistance), completing the circuit, allowing current to flow, and an orange light to be emitted. The current is extremely low, limited to about 500 microamperes by resistor R , and very little power is consumed.

A neon lamp is not shock proof or vibration proof (although more so than an incandescent), lasts for approximately ten years, and requires an additional light socket assembly for ease of replacement.

Because of the relatively high power consumed by an incandescent lamp, the relatively high voltage required by a neon lamp, and because of their mechanical sensitivity and limited longevity, both incandescent and neon lamps are being replaced and made obsolete by *light emitting diodes* (LEDs).

These newer devices are special semiconductor diodes with all the desireable features of semiconductors - low power dissipation, shock and vibration proof characteristics, and essentially infinite life. Light emitting diodes are discussed in detail in Volume Two - Part Two - Optoelectronics.

CRYSTALS

A **crystal** is a thin, solid slab of natural quartz material with molecules set in a regular geometric pattern or lattice structure. Its composition and form allows for unique circuit applications.

A particular type of quartz crystal, called a *piezoelectric crystal,* has exceptional electrical capabilities.

- When the crystal is subjected to an alternating mechanical pressure, positive and negative charges are set up on its opposite faces to generate an AC voltage . The frequency of the AC voltage depends on the crystal's size and type of cut and will be maintained at a constant value when used in an appropriate electronic circuit.

- When an AC voltage is applied across the faces of the crystal, it will contract and expand periodically and generate mechanical vibrations. As the frequency of the AC varies, the vibrations will change proportionately.

The *piezoelectric effect* was a phenomenon that was first noted in 1880 by **Pierre and Jacques Curie**, two French physicists, during their extensive study of crystalline materials. The name is derived from the Greek word *piezein,* which means to squeeze or to press.

APPLICATIONS

CRYSTAL-CONTROLLED OSCILLATOR

An *oscillator* is a circuit that generates an AC voltage at a specific fundamental frequency. In a typical oscillator that is not crystal-controlled, an inductor and capacitor are connected in parallel at the input of a circuit. An additional LC network, tuned to a harmonic or subharmonic of the LC network at the input, can be connected at the circuit output if a frequency other than the fundamental frequency is desired.

The circuit requires that a *positive* or *regenerative feedback* loop be connected between output and input to have the circuit oscillate. When the feedback portion of the amplified output voltage exceeds the voltage at the input that caused it, oscillation will occur.

An oscillator may be considered to be an amplifier with positive feedback and appropriate circuit parameters that are added to restrict the oscillations of the circuit to a single frequency.

When frequency-stability and precision are prime requirements, an oscillator is **crystal-controlled** with the piezoelectric crystal acting as the tuning element or resonant section of the oscillator input.

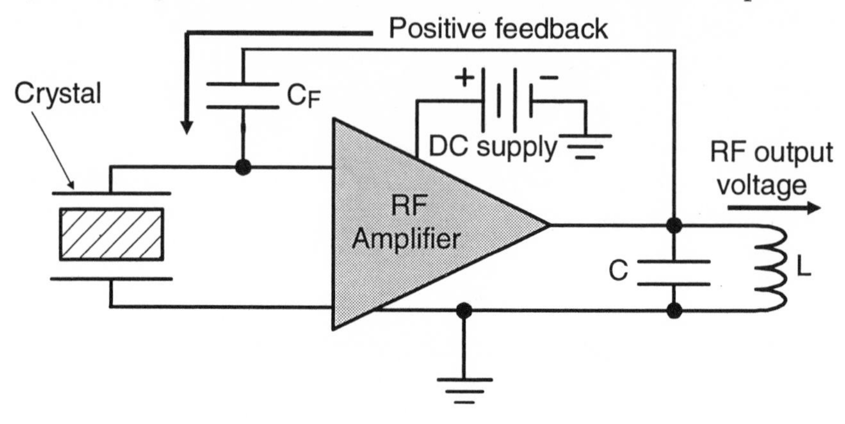

Crystal-controlled Oscillator Circuit
Figure 13.6

In the oscillator circuit of Figure 13.6, a piezoelectric crystal, rather than an LC network, acts as the input tuned circuit. The crystal will resonate at the specified radio frequency when a portion of the amplified RF at the output is coupled to the input, in phase with the input, through the feedback capacitor, C_F. A resistor in series with C_F can be used to reduce feedback when required.

With a crystal replacing an LC circuit, the oscillator is extremely stable, maintaining its frequency despite changes in the amplitude of the DC supply voltage or changes in circuit temperature.

In addition, the crystal has an extremely high quality factor (Q) and provides a very selective response to the desired frequency while sharply rejecting all other frequencies above and below the crystal's resonant frequency.

Variations of this circuit are used in radio and TV broadcast transmitters, other commercial and amateur communications equipment, and citizen's band (CB) radio to initiate and maintain the stable, constant frequency, AC voltage required for proper transmission and control of a communications system carrier signal.

CRYSTAL-CONTROLLED CLOCK

Every computer must have a clock as part of its circuitry to set its "rhythm", allowing it to operate in a properly timed manner and to precisely exchange timed data with its peripherals. The crystal provides the stability needed for the complex multi-phase clocks used in large computers. Crystals are also used to achieve high levels of timing accuracy in consumer quartz crystal watches and clocks.

AC FILTERS

Crystals can be used instead of less precise inductor/capacitor networks in AC filters (see Chapter 12 - Magnetic Components), allowing specific frequencies to be passed though a circuit while all other frequencies are sharply rejected. In this manner, they are applicable for high-quality, highly-stable AC filters in radio receivers, and other similar audio and radio frequency applications.

MICROPHONES, PHONOGRAPH CARTRIDGES, AND HEARING-AID PICKUPS

The electrical output of a crystal depends on the mechanical vibrations impinging on the crystal microphone, phonograph cartridge stylus, or hearing aid pickup used in these devices. Variations in the mechanical vibrations generate a corresponding audio frequency voltage; it is then amplified in its associated electronics section. The output of a crystal is considerably higher than that of a magnetic microphone or magnetic cartridge pickup.

CRYSTAL SPECIFICATIONS

FREQUENCY OF OPERATION
Ranges from 1 kHz to 1.5 gigahertz, however, the majority of crystals are made for operation in the megahertz region

TOLERANCE
Ranges from ± 0.001% (10 ppm) to ±0.005% (50 ppm) of the specified frequency

FREQUENCY-STABILITY
Specified from ± 0.0005% (5 ppm) to ± 0.01% (100 ppm) change over specified temperature range.

Oven controlled oscillators can maintain frequency stability as precise as ± 0.00002 % (0.2 ppm) over the specified temperature range of the oven.

LONG TERM FREQUENCY DRIFT (AGING)
Specified as drift per year at a specified temperature ranging from ± 0.0001% (1 ppm) to ± 0.0005% (5ppm).

MODE OF OPERATION
Tuning mode of subsequent resonant circuits - fundamental frequency, third harmonic, fifth harmonic, etc.

OPERATING TEMPERATURE RANGE
- Military and space: From -55°C to +125°C
- Industrial, commercial, and consumer: Typically, 0°C to +85°C

When extreme stability is required, a crystal-controlled oscillator is often installed in a thermostatically-controlled oven that maintains a fixed temperature (within the tolerance of the oven).

PACKAGING
A great variety of crystal holders are available with different dimensions and styles. Their selection depends on frequency, nature of the end use, operating temperature, and the required environment.

Plastic packages are used in commercial applications; hermetically-sealed metal and ceramic packages are used for military and space applications.

ENVIRONMENTAL REQUIREMENTS
As with all components, the ability of a crystal to resist the stresses of the environment will be dictated by the specific application for which it is intended.

REINFORCEMENT EXERCISE

Answer TRUE or FALSE

1. A connector is a mechanical coupling device that provides rapid joining or separation between sections of an electrical or electronic circuit or system.

2. Conductors are connected to the terminals of adjoining connecting sections by only soldering or wire-wrapping.

3. Most connectors are standardized with regard to their electrical and mechanical parameters, and there is interchangeability among most connectors, regardless of the manufacturer.

4. The contacts of high insertion force connectors have no contact wear or abrasion resulting in long life of the contact surfaces.

5. Zero insertion force (ZIF) connectors are used in test fixtures where connectors are engaged and disengaged frequently.

6. Connectors are considered to be the most reliable component in an electronics system.

7. Both incandescent and neon lamps can be energized by either AC or DC.

8. An incandescent lamp emits light that covers the light spectrum that includes non-visible infrared and visible light. The emitted radiation is in the form of white light.

9. A neon lamp generally has a longevity of about ten years and is used with low voltage (24 volts or lower) for proper operation.

10. Piezoelectric crystals will oscillate only if a DC voltage is applied across the faces of the crystal.

Answers to this reinforcement exercise are on page 282.

CHAPTER
FOURTEEN

TECHNOLOGY TRENDS FOR PASSIVE COMPONENTS

SURFACE MOUNTED DEVICE (SMD) TECHNOLOGY

COMPONENT MINIATURIZATION

USE OF ALLOYS TO REPLACE GOLD

TECHNOLOGY TRENDS FOR PASSIVE COMPONENTS

INTRODUCTION

The value of passive components has long established in the line-up of electronic components. Compared to the newer, and perhaps more dynamic devices called *semiconductors*, passive components are generally more readily interchangeable and more reliable than semiconductors. As a result, there have been very few changes in the basic structure of passive components during the past several decades. Passive component technology is a mature and stable science; this fact is reflected in the maturity and stability of the passive component industry.

Although passives have experienced relatively few changes, there have been a few important and innovative developments, specifically in the areas of component mounting techniques, component miniaturization, and in the use of alloys in place of gold.

SURFACE MOUNTED DEVICE (SMD) TECHNOLOGY

The most significant trend in passive component technology has been in the accelerating use of components being made available in a variety of forms for mounting onto the surfaces of printed circuit boards. This new package method is classified under the heading of Surface Mounted Device (SMD) technology.

Historically, component surface mounting began in 1971 when the Sony Corporation of Japan developed a packaging technique called *Metal ELectrode Face-bonding* (MELF), a process using small cylindrical resistor and capacitor pellets bonded to the surface of a PC board. Other Japanese companies soon developed a different approach by using rectangular carbon-film and metal-film resistor chips and ceramic capacitor chips.

Within two to three years of the onset of component surface mounting, several resistor and capacitor manufacturers in the United States and in Europe began producing similar resistor chips as well as capacitor chips of ceramic, mica, plastic film, glass, and tantalum. The proliferation of additional companies becoming involved in SMD technology has brought about its rapid expansion. This acceleration is expected to continue throughout the rest of this century and into the next.

SMD technology started with fixed resistors and capacitors (see Fig.14.1), it has expanded into the area of trimmers, potentiometers, switches, relays, inductors, transformers, and connectors. This new packaging approach includes semiconductors as well.

Resistor and Capacitor Chips for Surface Mounting
Figure 14.1

With the more complex, multi-terminal devices, such as trimmers, potentiometers, switches, relays, and transformers, many new, low profile, multi-leaded packages were designed for mounting to the surface of a PC board. These were small-outline packages with either gull-wing or J-bend leads. (See Figure 14.2)

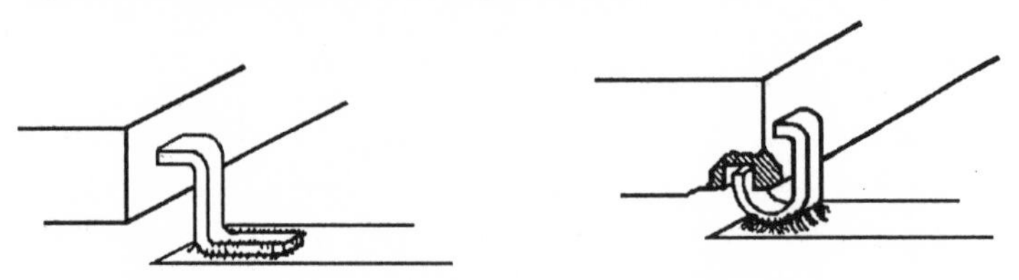

Gull-Wing Lead **J-Bend Lead**
Figure 14.2

Although it appears that the gull-wing leaded package is achieving greater popularity with component manufacturers in the United States, Japan, and Europe, the J-bend-leaded package provides an additional option that reduces the need for surface area. Some component manufacturers are moving in that direction.

SMD technology is having a major impact on PC board assembly and equipment manufacturing since it is quite different from traditional packaging techniques. The older approach uses leaded components mounted on one side of a PC board with leads inserted through predrilled holes to be wave-soldered on the underside of the board. This process is referred to as *through-hole technology*.

Assembly methods for surface-mounted devices require approaches that are different than those that have been used for automated assembly of the older leaded components.

SMD ASSEMBLY

Tape and reel bulk packaging is still the standard for automatic, high-speed, high-volume assembly. There are, however, variations of assembly processes that are unique to surface mounting device technology.

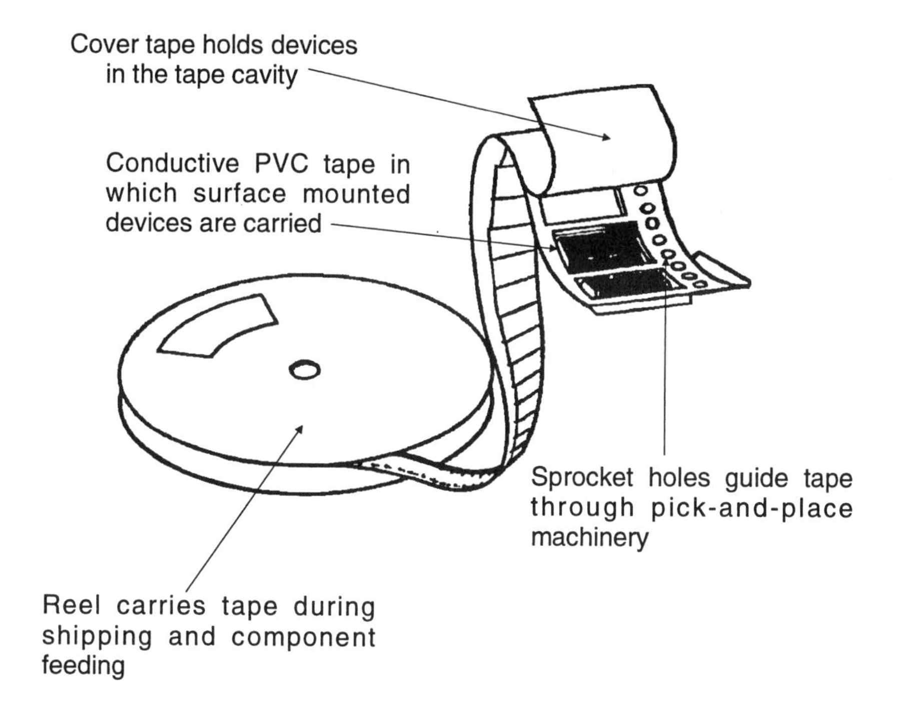
Cover tape holds devices in the tape cavity
Conductive PVC tape in which surface mounted devices are carried
Sprocket holes guide tape through pick-and-place machinery
Reel carries tape during shipping and component feeding

Tape and Reel Bulk Packaging
Figure 14.3

With SMD assembly:

- A solder-bearing paste is screen-printed onto the appropriate trace or solder-pad area of the board.
- A fixture on the automatic pick-and-place equipment removes the component from the tape and places it on the designated area of the PC board. (See Figure 14.4)

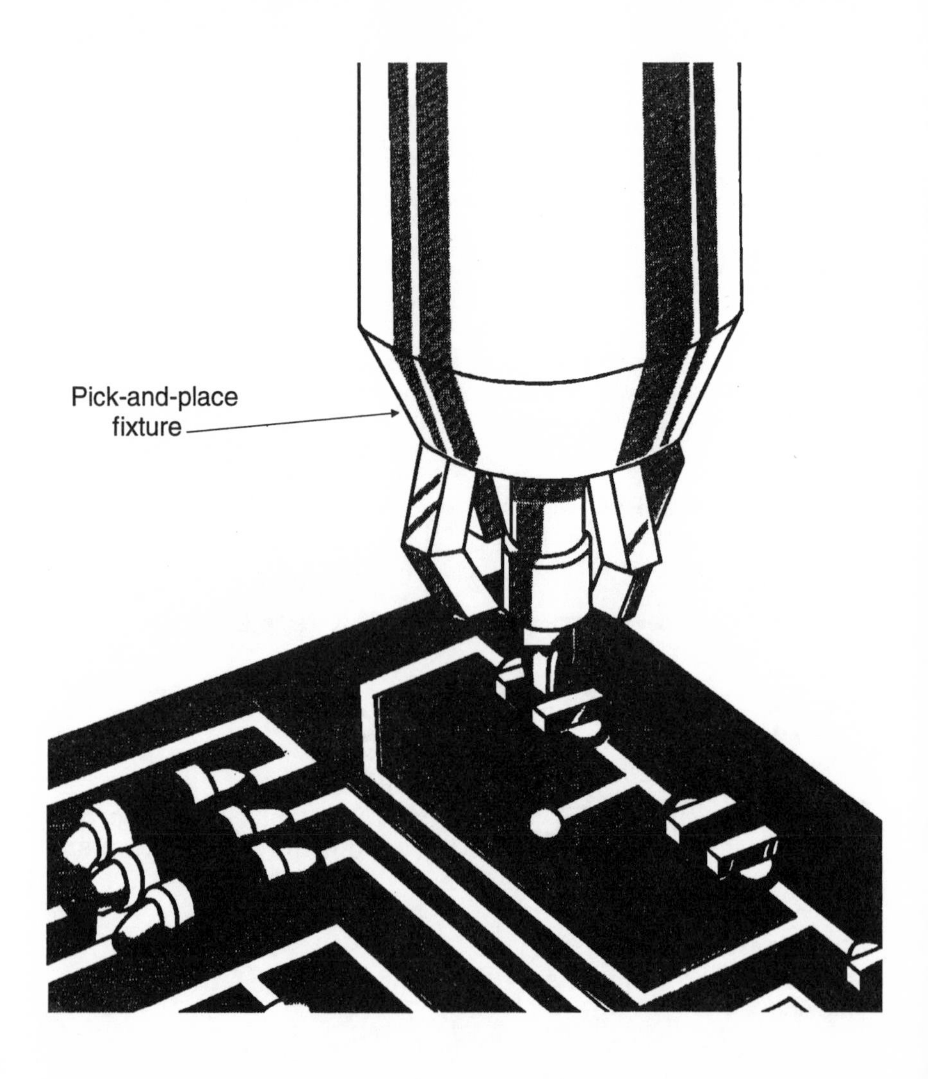

Placement of SMD on PC Board
Figure 14.4

- During this process, tiny drops of epoxy glue are applied to the undersurface of the component. The epoxy glue is hardened, temporarily holding the component until it is permanently soldered to the board.

- Instead of the older wave solder method, a vapor-phase reflow soldering process permanently bonds the component to the PC board.

- Vapor-phase reflow soldering is accomplished by heating vapors around the assembled board to melt the solder-bearing paste that was previously applied. The solder is applied more uniformly, at a lower, safer temperature and under more controlled conditions than with wave-soldering techniques.

Not only are resistors, capacitors, and small value inductor chips available in SMD form, in addition, switches, electromechanical relays, and crystals are being assembled in both J-bend lead and gull-wing lead packages for SMD applicatiions.

ADVANTAGES OF SMD TECHNOLOGY

Compared to through-hole techniques, SMD technology provides:

- Leadless component chips, pellets, or multi-leaded components in gull-winged or J-bend leaded packages that can be mounted on both sides of a PC board offering savings in PC board "real estate".

- These leadless component chips, pellets, and the smaller surface-mountable packages are lighter in weight than the older, traditional, leaded components.

- With smaller devices, shorter distances between components can be achieved, permitting shorter copper conducting strips (traces) on the PC board. Shorter traces reduce circuit capacitance and inductance, resulting in faster switching and higher-frequency circuit capability for improved overall circuit performance.

- Solder is applied more uniformly, under more controlled conditions and at a lower, safer temperature than that of the wave-solder technique inherent in traditional through-hole PC boards.

- Cost savings in PC board manufacture and assembly can be realized by eliminating holes for component leads in the boards and the forming and cutting of the leads on component packages.

- Surface mounted components are less susceptible to radiated noise or interference than long wire leads on axial-leaded components. Long wire leads sometimes act as antennas to radiate and receive high frequency noise.

Initially, like any new technology, there are several inherent disadvantages that work against the immediate and universal acceptance of an SMD approach.

DISADVANTAGES OF SMD TECHNOLOGY

- SMD requires costly automatic placement (pick and place) equipment and special soldering and test equipment. To justify these expenditures, SMD technology must be used for high volume production as in the automotive, radio and TV, consumer communications equipment, and low-priced computer industry.

  Several companies have begun SMD foundries to service the needs of equipment manufacturers with small production runs, particularly in the production of military and space equipment.

- In the late 1980s, about 45% to 55% of standard leaded devices were available in SMD form. From 10% to 20% of all passive components are currently unavailable in SMD form.

- The difference in the thermal expansion of the component chip and the PC board can result in fatigue failure of a solder joint after relatively few thermal cycles and can cause the chip to detach from the PC board.

  One approach to this problem is to use a conduction-cooled PC board designed specifically for surface mounted components.

  With multi-leaded components mounted in a gull-winged or J-bend leaded package, the difference in thermal expansion presents no problem since the flexible leads of the package absorb the differences in expansion.

- With components mounted on both sides of a PC board, testing procedures are more complex. A special test fixture resembling a clam-shell with test points on both inner halves is generally used for electrical testing. Test probe pins for SMD boards are much smaller and must be carefully positioned. Circuit test pads are generally made part of the artwork for the PC boards.

- While surface mounted devices can be cost-effective in large volume production, at the present time, small-quantity cost of SMD components remains higher than their leaded counterparts.

- A variety of chip sizes and non-standard dimensioned surface-mounted, multi-leaded packages exist throughout the industry. Because of this situation, it is difficult to design a circuit layout while maintaining a sense of confidence that there will be long-run availability of the particular component chosen.

- The confusing array of existing and proposed SMD standards are inhibiting faster growth and acceptance. There are at least 13 different package standards, each with its own dimensions.

COMPONENT MINIATURIZATION

The manufacture of small low-current, low-voltage passive components, particularly for use in PC board surface mounting, has significantly increased.

Primarily, their function is for interfacing with the low-current, low-voltage requirements of high density integrated circuits being used in computers, telecommunications systems, home appliances, automotive circuits, and automatic testing and measuring equipment (ATE).

Dramatic improvements in plastic and other insulating materials are making a new generation of miniature and sub-miniature components possible and include:

DIP and membrane switches
Resistors
Electromechanical relays
Trimmers
Inductors
Capacitors
Potentiometers
Connectors

USE OF ALLOYS TO REPLACE GOLD

The traditional use of gold plating on the contacts of switches, relays, and connectors has escalated the cost of some of these components beyond a practical level. In evaluating these costs, many manufacturers have questioned the justification of the extensive use of gold, silver, platinum, and palladium in previously routinely applied plating.

When these metals were relatively inexpensive, contacts were generally heavily plated with gold and other precious metals to achieve low contact resistance and preventation of corrosion for proper circuit performance under dry circuit and severe environmental conditions. After extensive research and development of new alloyed materials, manufacturers have found that gold and some of the other expensive metals can be replaced with these alloys without compromising the quality of performance.

In addition to the use of alloys, a trend has developed toward selectively goldplating the contacts of switches, relays, and connectors where complete elimination of precious metal is impossible. There has been a strong decline in the use of these costly materials while, at the same time, performance characteristics have not been sacrificed. These efforts have been instrumental in maintaining low-cost, high-quality components.

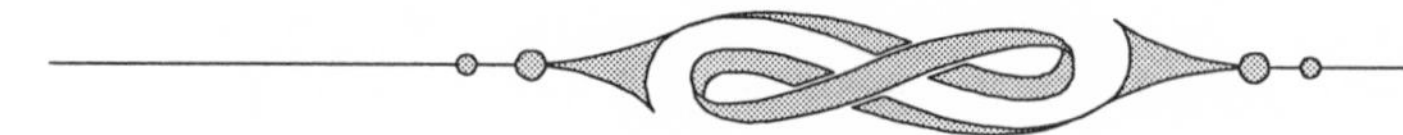

Part II of this volume has covered the subject of passive components used in electrical and electronic circuitry. These components represent a significant section of the overall aspect of electronic circuits and systems.

In Volume Two - Discrete Semiconductors and Optoelectronics, semiconductors, the active components, will be examined in detail.

Semiconductors, used in conjunction with passive components, provide the means to implement the design and construction of the modern electronic circuits and systems that have become an integral part of today's electronics environment.

APPENDIX

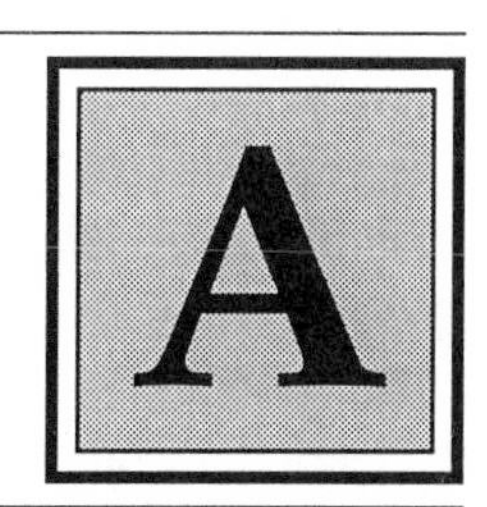

ANSWERS TO REINFORCEMENT EXERCISES

RESISTORS

CAPACITORS

SWITCHES, KEYBOARDS, AND ELECTROMECHANICAL RELAYS

MAGNETIC COMPONENTS

MISCELLANEOUS PASSIVE COMPONENTS
- CONNECTORS
- INDICATOR LAMPS
- CRYSTALS

ANSWERS TO REINFORCEMENT EXERCISES

RESISTORS

Questions are listed on pages 159 - 160.

1. True

2. True

3. True

4. True

5. True

6. True

7. False - A thermistor is a special resistive component with a clearly specified temperature coefficient (TC). The resistance of a positive temperature coefficient (PTC) type increases as temperature rises.

   Some PTC types are initially very low in resistance and remain at a low value with a relatively small increase in temperature. At the critical (transition) temperature, the resistance increases very suddenly. This characteristic makes the device behave as a resettable fuse or a normally ON switch that is actuated by increasing temperature.

   The resistance of an NTC thermistor will decrease as its temperature increases. This characteristic allows the thermistor to act as a temperature compensating device for a PTC section of a circuit.

8. False - Thin film technology is used in the manufacture of resistor chips to provide a lower TC and about ten times closer temperature tracking than thick film.

9. False - Multi-turn potentiometers and trimmers are designed for more precision or improved resolution than single-turn devices.

10. True

11. True

12. True

Resistor identification - Using the color code

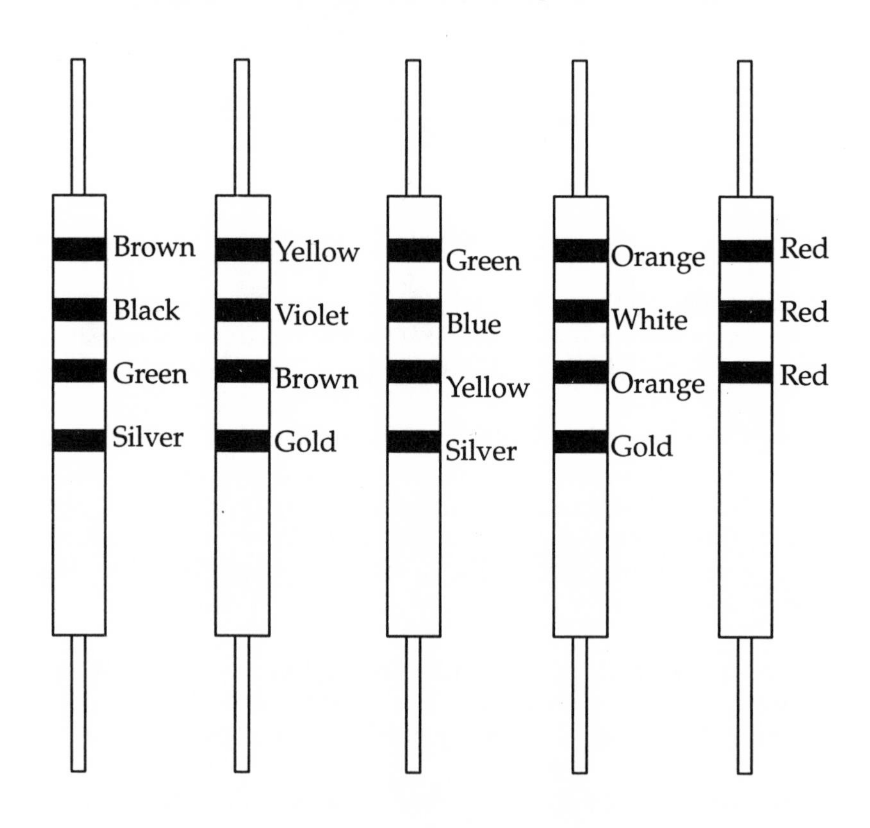

Values of the resistors:

1 MΩ	470Ω	560kΩ	39kΩ	2200Ω
±10%	±5%	±10%	±5%	±20%

CAPACITORS

Questions listed on pages 185 - 186.

1. True

2. False - The capacitor package seal and the external material of the package are the determining factors in how effectively the capacitor resists the effects of environmental stresses, other than temperature.

3. False - The purpose of manufacturing a polarized capacitor is to provide a large capacitance in a physically small package. Nonpolarized types can be made with large capacitance, but their physical size may be several times larger than the comparable polarized types.

4. True

5. False - Tantalum oxide capacitors are preferred because of their more favorable characteristics.

6. True

7. True - In this case, the polarized capacitor will always have the proper voltage polarity applied across its terminals despite the variations in voltage across the capacitor. The voltage varies because of the amplitude of the positive and negative half cycles of the AC voltage.

8. True

9. True

10. True

11. True

12. False - A high resistance load across a capacitor produces a long discharge time (large R = long time constant). Zero resistance across the charged capacitor (zero time constant) will act to discharge its voltage instantaneously.

13. True

14. False - The capacitive reactance, or AC resistance, of a capacitor is equal to one divided by $2\pi f C$.

$$\text{Capacitive Reactance} = X_C = \frac{1}{2\pi f C}$$

As frequency increases, capacitive reactance decreases.

15. True

16. True - Since the spacing between chips mounted on a hybrid IC substrate is much smaller than the spacing between components hard-wired to a PC board, the speed and frequency response may be improved.

17. True

18. True

MECHANICAL SWITCHES, KEYBOARDS, AND ELECTROMECHANICAL RELAYS

Questions are listed on pages 217 - 218.

1. True

2. True

3. True

4. False - The switching action of a magnetic switch is accomplished by either placing a magnet near its contacts or by moving the magnet away from its contacts.

5. False - As with any mechanical system, there is a tendency for an electromechanical relay to wear down because of the effects of friction, material fatigue, and natural forces of erosion.

6. False - The purpose of a bounceless keyboard is to ensure that only one encoded character is generated when a key on the keyboard is depressed. A bounceless keyboard will not produce an additional, uncalled-for character.

7. True

8. True

9. False - A relay can be actuated with the application of either AC or DC voltage, but must be specifically wound for either type, or both.

10. False - A dry circuit is one in which the open-circuit voltage is less than 50 millivolts and in which currents of only a few milliamperes are flowing when the contacts are closed. Because of this low voltage and low current, oxides, sulfides, and other films can build up on contacting surfaces. These contaminants will introduce resistance between the contacts, effectively causing contacts to be open when they should be closed.

11. True

12. True

13. False - The contacts of the mercury-wetted relay are temporarily coated (not permanently plated) by a film of mercury through capillary action to overcome the effects of contact bounce and to maintain the conductivity of the contacts.

14. False - Mercury-wetted relays are mounted in a vertical or near-vertical position for proper operation.

15. False - Contact bounce can produce transients in a circuit that might result in circuit error or instability and must be eliminated or carefully considered in the circuit design.

16. True

IDENTIFICATION OF CONTACT CONFIGURATIONS

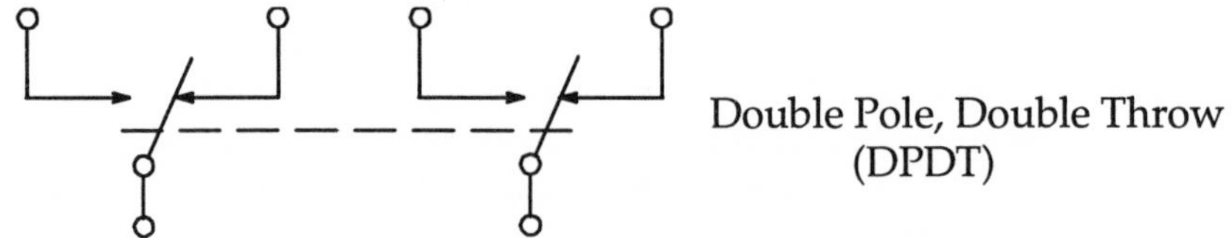

Double Pole, Double Throw
(DPDT)

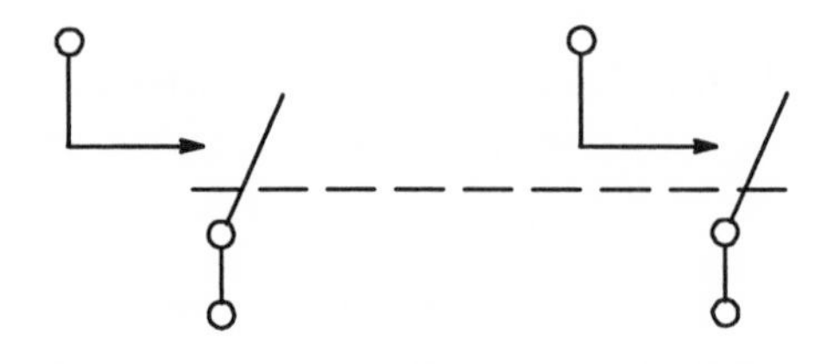

Double Pole, Single Throw
(DPST)
Normally Open (NO)

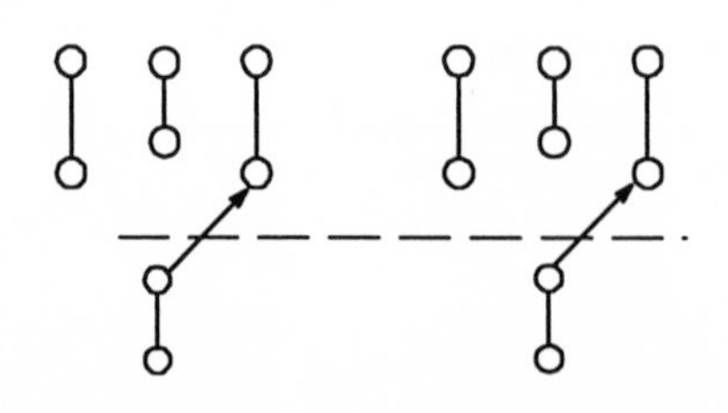

Double Pole, 3 Position
(DP 3P)

MAGNETIC COMPONENTS

Questions are listed on pages 243 - 244.

1. True
2. True
3. True
4. False - As frequency increases, inductive reactance increases, since inductive reactance (X_L) = $2\pi f L$.
5. True
6. False - Ferrite core inductors are used at frequencies between 20 kHz and 10 MHz.
7. True
8. False - RF chokes are used specifically for isolating high frequency signals from the supply voltage.
9. False - Secondary voltage varies proportionately to the turns ratio; current varies inversely to the turns ratio.
10. True
11. True
12. False - A transformer cannot couple (transfer) steady-state DC. It requires a changing magnetic field to function properly.
13. True
14. True
15. True
16. False - At least ten times the rated AC voltage is applied for one minute in a hi-pot test.
17. True

MISCELLANEOUS PASSIVE COMPONENTS

Questions are listed on page 261.

1. True

2. False - Conductors can also be connected to the connector terminals by crimping or through the use of screw terminals.

3. False

4. False - High insertion force connectors produce abrasion during insertion and removal resulting in reduced contact surface life.

5. True

6. False - Connectors are often considered to be the least reliable link in the electronics chain. At high frequencies, a connector can introduce increased resistance, inductance, and capacitance that might cause a mismatch between circuits. In addition, vibration and temperature cycling, as well as moisture, dust, and corrosive vapors can disrupt circuit operation by adversely affecting the proper mating of the connector contacts.

7. True

8. True

9. False - A neon lamp indicator lasts for ten to twenty years. At least 60 volts must be applied across its terminals to ionize, or break down, the neon gas which will then allow current to flow and cause emission of light.

10. False - Oscillation is produced by an amplifier having positive feedback from its output to its input. Because of the piezoelectric crystal's mechanical/electrical relationship, it acts as a highly-stable, tuned circuit, maintaining a precisely controlled frequency of operation.

    Simultaneously, when an AC voltage is applied across the faces of the crystal, it will contract and expand periodically and set up mechanical vibrations at the frequency of the AC voltage.

APPENDIX

GLOSSARY OF POPULAR ELECTRONIC TERMS

RESISTORS

CAPACITORS

MECHANICAL SWITCHES, KEYBOARDS, AND ELECTROMECHANICAL RELAYS

MAGNETIC COMPONENTS

CONNECTORS

RESISTORS

ADJUSTABILITY - The feature that allows the output of a potentiometer or the value of a rheostat to be set to a specific value.

AMBIENT TEMPERATURE - The temperature of the surrounding air or a surface to which a component is attached.

ATTENUATOR - A resistive network offering reduction of the amplitude of an electrical signal without any significant phase shift or frequency distortion.

BULK RESISTOR - A resistive component made by providing ohmic contacts between two points of a homogeneous, uniformly doped material.

CARBON COMPOSITION RESISTOR - A fixed resistive component produced by combining carbon with a filler and binder. After bonding leads to the ends of the resistor body, it is then encapsulated in a plastic jacket.

CARBON FILM RESISTOR - A fixed resistive component formed by depositing a thin carbon film on a ceramic core. Leads are bonded to the ends of the carbon film coated core (body) and an epoxy coat is applied around the body. Also called DEPOSITED CARBON.

CERMET - A conductive mixture used in making film resistive elements. The first half of the term is derived from ceramic, the second half from metal.

CHIP RESISTOR - A small, rectangular resistor chip used as part of a hybrid integrated circuit and available in either thick film or thin film construction. Thick film resistor material is screened onto a ceramic or glass substrate; thin film is vacuum-deposited. Resistance is adjusted by scribing, sandblasting, or trimming with a laser beam.

COLOR CODING - The standard Electronic Industries Association (EIA) code that designates the resistance value and tolerance of a fixed resistor. Color bands are painted on one end of the resistor body to provide this information.

CONDUCTIVITY - The ability of a material to conduct or transmit electrical current or heat. It is the reciprocal of resistance (1/R) and is measured in mhos (ohm spelled backward). Conductivity is designated with the symbol, ℧.

CONTACT RESISTANCE - When a contact touches the resistive element in a potentiometer, rheostat, or trimmer, a variation can be noted in the electrical output that is not present in the electrical input. It is measured as an equivalent resistance between the contact and element, expressed in ohms, or as a percentage of the total resistance.

DEPOSITED CARBON RESISTOR - See CARBON FILM RESISTOR.

DERATING - At ambient temperatures higher than the maximum specified for full-rated power, a reduced amount of power is allowable in the resistor. The allowable amount is shown in derating curves that normally end at zero power at the maximum surface temperature applicable to the resistor type. Additional derating may be needed for high-reliability applications, extremely high altitudes, or in the case of chassis-mount types, reduced mounting area.

DIELECTRIC STRENGTH - The maximum voltage of the dielectric (insulation) of a resistor, potentiometer, or trimmer which is applied between the case and all terminals that are connected to each other without exceeding a specified leakage current. This specification also applies to all other electrical components.

ELECTRICAL ROTATION - The total rotation of the shaft in a potentiometer, rheostat, or trimmer over which electrical continuity is maintained between the contact and the resistive element. Electrical rotation includes the flat areas. See FLAT.

END RESISTANCE (Re) - That portion of a potentiometer, rheostat, or trimmer that remains unavailable when the wiper has reached the end of its travel.

FILM RESISTOR - A fixed resistor that is characterized by the film properties of its resistance material rather than its bulk properties. It consists of a thin layer of resistive material deposited on an insulated form.

FLAT - The area that might possibly exist at each end of the element of a potentiometer, rheostat, or trimmer and produces no change in output voltage (or resistance) with further rotation of the shaft.

FLIP-CHIP RESISTOR - An unencapsulated resistor chip on which bead-type leads terminate on one face to permit "flip" (face-down) mounting of the resistor chip through contact of the terminals with interconnective circuitry.

GIGOHM - A thousand megohms (10^9 ohms) or a billion ohms.

HEAT RISE - The self-generated temperature increase of a resistor in during circuit operation. The total surface temperature is equal to heat rise plus ambient temperature. This definition applies to all other components as well.

INDUCTANCE (L) - The property of a wire-wound or spiral-cut film resistor which causes AC resistance to change with changing frequency. Inductance is effective only in AC circuits and has no effect in DC circuits.

INSERTION LOSS - The effect of an attenuator set at a minimum resistance on the load current of a circuit. It is the ratio of the current with the attenuator in the circuit to the current that would exist if it were not present. It is expressed as a power loss in decibels.

KILOHM - One thousand (10^3) ohms.

LINEARITY - The relationship of resistance, or resistance ratio, to the angular or linear travel of a variable resistor.

LOCATING LUG - An extension on the front of a potentiometer or rheostat which is inserted through the mounting panel to prevent rotation of the body of the control.

MAXIMUM WORKING VOLTAGE - The maximum specified voltage that may be applied across a resistor.

MEGOHM - One million (10^6) ohms.

METAL-FILM RESISTOR - A resistor formed by depositing an extremely thin layer of metal alloy on a ceramic or glass rod. After leads are bonded to the body, it is then encased in a protective shell.

METAL-OXIDE RESISTOR - A type of film resistor in which tin oxide is the resistance material deposited on a substrate. The tin oxide provides a low temperature coefficient value and high stability.

MILLIOHM - One thousandth (10^{-3}) of an ohm.

MULTI-TURN POTENTIOMETER - A potentiometer that requires more than one complete (360°) rotation of its control shaft for the slider (moving contact) to traverse the entire length of the resistance element. Output voltage depends on the shaft position and the number of turns of the potentiometer.

NOISE - Unwanted voltage fluctuation generated within a resistor. Total noise of a resistor is a factor of the resistance material, resistance value, temperature, and current flow. It can also be caused by loose end caps, loose leads, or cracked bodies. For variable resistors, noise may also be caused during the movement of the contact over turns of wire, or an imperfect electrical path between the contact and resistance element.

NON-WIREWOUND TRIMMER - A trimmer characterized by the unbroken nature of the surface area of the resistance element to be contacted. Contact is maintained over a continuous, unbroken path. The resistance is achieved with materials other than wire, such as: carbon, conductive plastic, metal film, and cermet.

NON-INDUCTIVE WIREWOUND RESISTOR - A resistor wire wound around a core which causes the magnetic field of one turn to cancel the magnetic field of an adjacent turn. The wire is coated with insulation, folded in half, and then both halves wound around an appropriate core. It is then insulated with terminals attached.

OHM (Ω) - The unit of resistance, named after Georg Simon Ohm, a 19th century German physicist.

POTENTIOMETER - A three-terminal component having a terminal connected to each end of a resistance element, with the third terminal connected to a moving contact or wiper.

POWER RATING - The maximum specified power dissipated in a resistor under specified conditions of mounting and environment.

RESISTANCE -The specific property of a component or conductor which acts to oppose an applied voltage and determines the current flowing in a circuit. Resistance depends on the material's molecular structure, size, and temperature. Resistance is measured in ohms.

RESISTANCE ELEMENT - A continuous, unbroken length of resistive material without joints, bonds, or welds, except at the termination of the element (the electrical terminals connected to each end of the element), or at an intermediate point, such as a center-tap.

RESISTANCE TOLERANCE - The permissible percent deviation from the nominal resistance value specified by the manufacturer at standard or stated environmental conditions.

RESISTOR - A basic component that provides resistance in electrical and electronic circuits. In a circuit having an applied voltage, a current flow results that is determined by the voltage and the value of the resistor, in accordance with Ohm's Law.

RESISTOR NETWORK - A group of individual resistors that can be arranged in a multitude of connection options in a single package. The resistor elements can be interconnected internally, providing various types of circuit configurations, or can be supplied as individual, unconnected resistors.

RESOLUTION - The smallest possible incremental change of output in a variable resistor.

RHEOSTAT - A two-terminal variable resistor with one terminal connected to one end of a resistance element and the other connected to its wiper. A potentiometer that has three terminals can be made to act as a rheostat by using only two of its three terminals (either end-terminal and the wiper terminal).

SINGLE-TURN POTENTIOMETER - A potentiometer requiring 360°, or less, to enable the wiper to traverse the total resistance element.

TAPER - Part of a potentiometer specification that indicates the resistance change between the control (adjustable) terminal and either end terminal as the potentiometer actuator is moved.

THERMISTOR - Derived from "thermal resistor", it is a device with a fixed and precisely specified temperature coefficient. It can have either a positive temperature coefficient (PTC) or a negative temperature coefficient (NTC). Generally, PTC types act as resettable fuses; NTC types compensate for the changing temperature within a circuit to achieve a zero TC effect.

THICK FILM RESISTOR CHIP - Resistor elements made by a mixture of ceramic and various metals (cermet), silk-screened onto an insulating material (substrate), and then fused at high temperature to form a resistor chip. Conductive electrodes of palladium-silver are then fused on each end of the chip body. Cermet is most often made from ruthenium-dioxide.

Resistor chips made with this screen-printing process are manufactured for commercial, industrial, and consumer hybrid integrated circuits where tolerance and temperature effects are not critical. These chips are designed with surface mounted device (SMD) packaging techniques that provide definite advantages in component assembly and space utilization.

THIN FILM RESISTOR CHIP - A resistor element made by a vapor-deposition of a very thin metallic film onto a ceramic or glass substrate. The metallic film is either chromium-cobalt, nichrome, or tantalum, with interconnecting conductors on the substrate which are usually made of gold.

Thin film resistors provide the characteristics needed for hybrid integrated circuits designed for military and space equipment. They offer a low temperature coefficient, tight tolerance, a wide operating temperature range, and a high degree of reliability.

TRIMMER - A potentiometer or rheostat used for internal circuit adjustment or circuit calibration. Normally, a trimmer is not accessible to the end user but is meant for circuit adjustment in the final stages of production or for periodic circuit maintenance.

VARISTOR - A metal (zinc) oxide device with variable resistance characteristics that act to protect against lightning and other excessive voltage surges.

Resistance of the device varies as a function of applied voltage. Below its rated voltage, its extremely high resistance has no effect on a circuit. Above its rated voltage, the varistor suddenly changes to an extremely low resistance value, acting to clamp the applied voltage to a safe level of operation.

VOLTAGE DIVIDER - A group of resistors connected in series across a voltage source. Terminals are connected at the junction points of the divider to provide a means of accessing fraction(s) of the available voltage source.

WIPER - The part of a variable resistor which is in contact with its resistive element allowing the output to be varied as the control is varied mechanically.

WIREWOUND RESISTOR - A fixed resistor in which the resistive element is a length of high-resistivity wire or ribbon wound onto an insulating core. It is then encapsulated in vitreous enamel, silicone, cement compound, or other insulating material.

CAPACITORS

AIR CAPACITOR - A variable capacitor with air as the dielectric material between the conducting plates.

ALUMINUM ELECTROLYTIC CAPACITOR - A capacitor with two aluminum electrodes that have an oxide film on the anode. The electrodes are separated by layers of absorbent paper saturated with a conductive electrolyte.

ANODE - The positive electrode of a polarized capacitor.

AXIAL-LEADED CAPACITOR - A capacitor whose leads extend from its ends along the axis of its body.

BLOCKING CAPACITOR - A capacitor used for blocking steady-state DC by introducing a very high resistance to DC while offering little resistance to an AC voltage.

BYPASS CAPACITOR - A capacitor with very low resistance to high frequency AC voltages which detours RF signals around a circuit element, generally, a DC power supply.

CAPACITANCE - The property of a capacitor which permits electrical charges to be stored in its dielectric when a potential difference exists between its plates or electrodes. The potential difference depends on the voltage applied to the capacitor's terminals. Farad (F) is the unit of measurement.

CAPACITANCE TOLERANCE - A manufacturer's guaranteed maximum deviation from the specified nominal capacitance at specified environmental conditions. Tolerance is expressed as a percentage of the capacitor's nominal value.

CAPACITIVE COUPLING - The transfer of an AC voltage from the output of one circuit to the input of another by means of a capacitor. Also called ELECTROSTATIC COUPLING.

CAPACITIVE REACTANCE (X_C) - The AC resistance produced by a capacitor and measured in ohms. X_C is equal to $1/2\pi fC$.

CAPACITOR - A component that consists of two conducting surfaces separated by an insulating material (dielectric), such as: air, paper, mica, plastic, ceramic, or metallic oxides (aluminum oxide and tantalum oxide). The unit of measurement of a capacitor is a farad (F). A capacitor is capable of storing electrical energy; the amount of energy stored is a function of its capacitance and the applied voltage. A capacitor can block DC and permit the flow of AC depending on its capacitance and the frequency of the applied voltage.

CAPACITOR-INPUT FILTER - A power supply filter used with a series resistance to change pulsating DC voltage to steady-state DC voltage.

CATHODE - A polarized capacitor's negative electrode.

CERAMIC CAPACITOR - A capacitor that has its dielectric made of a ceramic material, usually a form of barium-titanate.

CHARGE - The quantity of electrical energy stored in a capacitor.

COUPLING CAPACITOR - A component that acts as the output capacitor of one circuit and the input capacitor of another and serves to couple (transfer) AC voltage from one circuit to another.

CV PRODUCT - The capacitance of a capacitor, in farads, multiplied by its rated voltage.

DIELECTRIC - The insulating material (non-conductor) sandwiched between the two conducting plates of a capacitor. Typical dielectrics are air, paper, plastic, glass, ceramic, mica, and metallic oxides. A vacuum is an excellent dielectric and is the standard to which all other dielectrics are compared.

DIELECTRIC ABSORPTION - A measure of the failure of a capacitor to discharge completely. After a fully charged capacitor is momentarily discharged, the remaining charge is expressed as a percentage of the original charge. Dielectric absorption is affected by discharge time, voltage, and temperature. This parameter varies with different types of capacitors.

DIELECTRIC BREAKDOWN VOLTAGE - The voltage across the capacitor plates at which breakdown (rupture of the dielectric) occurs under specified test conditions. This value is generally higher than the rated breakdown voltage, the difference being the safety factor set by the manufacturer.

DIELECTRIC CONSTANT - Property of a dielectric material that determines how much electrostatic energy can be stored per unit volume when a unit voltage is applied across the plates. It is the ratio of the capacitance of a capacitor (with a given dielectric) to that of a capacitor having a vacuum dielectric.

DISK CAPACITOR - A small, disk-shaped, single-layer or multilayer ceramic capacitor, consisting of one or more dielectric insulators with opposing conductive surfaces. It is generally used for bypass applications.

DISSIPATION FACTOR (DF) - The ratio of the effective series resistance of a capacitor to its reactance and is measured in percent.

ELECTRODES - The conductive plates of a capacitor.

ELECTROLYTE - Current-conducting medium (liquid or solid) between two electrodes or plates of a capacitor, one or both of which are covered with a dielectric film.

ELECTROLYTIC CAPACITOR - A capacitor consisting of two conducting electrodes whose anode has an aluminum or tantalum oxide film on it. The film acts as a dielectric or insulating material. Nonpolarized electrolytic capacitors generally have two similar anodes as well as a liquid cathode.

ELECTROSTATIC COUPLING - See CAPACITIVE COUPLING.

EQUIVALENT SERIES RESISTANCE (ESR) - The internal DC resistances of a capacitor. For purposes of calculation, it is treated as if it were an internal resistor concentrated (lumped) at one part and connected in series with the capacitor terminal.

FARAD (F) - The unit of measurement of capacitance named after Michael Faraday, a 19th century British physicist/mathematician.

FEEDTHROUGH CAPACITOR - A component that provides a desired value of capacitance between a conductor and the metal chassis or panel to which the capacitor is mounted. Used mainly for bypass applications at high frequencies.

FISSURES - Cracks in ceramic capacitors of varying severity which can cause capacitor failure. Fissures are most often caused by thermal shock. Some extremely small fissures may not cause immediate capacitor failure but could do so after long periods of operation.

FIXED CAPACITOR - A capacitor with a specified capacitance that cannot be varied.

IMPREGNANT - A material, usually liquid, used to saturate a paper dielectric to replace the air between its fibers. The impregnant increases the dielectric strength and dielectric constant of the capacitor.

LEAKAGE CURRENT - An extremely small value of direct current that flows through a capacitor after it is charged.

LEAKAGE RESISTANCE - The ratio of the DC voltage applied to the terminals of a capacitor and the resulting leakage current flowing through the dielectric. This leakage current exists with either AC or DC voltage. A minimum value of leakage resistance is specified on a data sheet typically as several thousand megohms.

LIQUID-FILLED CAPACITOR - A capacitor in which a liquid impregnant occupies substantially all of the case volume not taken by the capacitor elements and their connections. Space is allowed for expansion of the liquid under changing temperature conditions.

LOAD LIFE - The rated life of a capacitor at its maximum rated condition of voltage, temperature, and ripple current. Load life is usually expressed in thousands of hours.

METALLIZED CAPACITOR - A self-healing, fixed capacitor. A thin film of metal is vacuum-deposited directly on a paper or plastic dielectric. When a breakdown occurs, the metal film around it immediately vaporizes.

MONOLITHIC CERAMIC CAPACITOR - A single block of material consisting of multiple layers of ceramic dielectric insulators separated by connected metal electrodes. Also called MULTILAYER CERAMIC CAPACITOR (MLC).

NPO - An ultra-stable capacitor with a very low temperature coefficient (30 ppm/°C) The term is derived from the phrase "negative-positive-zero".

PAPER CAPACITOR - A fixed capacitor consisting of two strips of metal foil separated by oiled or waxed paper rolled together into a compact wad. The foil strips can be staggered so that one strip projects from each end, with tabs generally added. The connecting wires are attached to the strips or tabs.

POWER FACTOR (PF) - A measure of the overall efficiency of a capacitor. Power factor ratings take all internal losses into consideration in a complex formula. This parameter, for some capacitors, is often very important in AC applications, such as motor-starting and motor-running.

RADIAL-LEADED CAPACITOR - A capacitor whose leads radiate from its body. See AXIAL-LEADED CAPACITOR.

RATED SURGE CURRENT - The maximum charging or discharging current that is permitted without damaging the dielectric and is specified by the manufacturer.

RATED VOLTAGE - The maximum DC voltage that can be safely applied across the terminals of a capacitor at a specified temperature without causing dielectric breakdown. Rated voltage is specified by the manufacturer.
See WORKING VOLTAGE.

RIPPLE CURRENT OR VOLTAGE - The varying part of a steady-state DC current or voltage. This value is generally very small compared to the magnitude of the steady-state DC voltage.

ROTARY CAPACITOR - A capacitor whose value can be changed by rotating the shaft of a variable capacitor, thereby changing the position of one set of conducting plates with respect to the other without changing the distance between them. This type of capacitor is used for adjusting the frequency of a tuned or resonant circuit.
See VARIABLE CAPACITOR.

SILVERING - The process by which metal terminals are applied to the ends of a ceramic chip and then fired to the ceramic body.

SINTERING - The process by which metals or powders are bonded by first pressing them into a desired shape. They are then heated at a high temperature to form a strong, cohesive body.

SOLID TANTALUM CAPACITOR - A polarized capacitor with a solid tantalum oxide dielectric instead of a liquid. A sintered pellet is used for high capacitance values and a wire anode for lower value capacitances. Also referred to as a SOLID-ELECTROLYTE TANTALUM CAPACITOR.

SURGE CURRENT - The instantaneous current entering (charging) an uncharged capacitor when a voltage is applied across its terminals or, the instantaneous current leaving (discharging) a charged capacitor when a load is connected across the capacitor.

TIME-CONSTANT - The length of time, measured in seconds, that it takes a capacitor-resistor charging circuit (RC network) for the capacitor to reach 63% of the applied voltage. In an RC discharge circuit, it is the interval of time for the capacitor voltage to reduce to 37% of the voltage of the fully charged capacitor. A time-constant (in seconds) is equal to the product of the capacitance (in farads) and the resistance (in ohms).

TRIMMER - A variable capacitor of relatively low value connected in parallel with a fixed capacitor having a higher value. Acts to effectively increase the capacitance of the fixed capacitor for fine-tuning a circuit.

TUNED OR RESONANT CIRCUIT - A circuit that is selective in either passing or rejecting a specific frequency or group of frequencies. It generally consists of an inductor and a capacitor connected in either series or parallel. The inductor or the capacitor can be adjusted to achieve resonance at a desired frequency. This is the frequency at which the inductive reactance (X_L) is equal to the capacitive reactance (X_C).

TUNING CAPACITOR - A variable capacitor used for adjusting the frequency of a tuned or resonant circuit.
See TUNED CIRCUIT.

VARIABLE CAPACITOR - A capacitor whose value can be changed by varying the facing area of its plates, as in a rotary capacitor, or by altering the distance between them, as in a trimmer capacitor.

WET SLUG TANTALUM CAPACITOR - A polarized capacitor whose cathode is a liquid electrolyte and its anode is tantalum with a tantalum oxide dielectric. These capacitors have the highest capacitance per unit volume, lowest DC leakage, and long-lasting shelf life.

WORKING VOLTAGE - The maximum DC voltage that can be applied to the terminals of a capacitor for continuous operation at the maximum rated temperature.
See RATED VOLTAGE.

MECHANICAL SWITCHES, KEYBOARDS, AND ELECTROMECHANICAL RELAYS

ABNORMAL BRIDGING - In a switch or relay, the undesired closing of open contacts resulting from contact bounce, a metallic bridge, or a protrusion caused by arcing at the contacts.

ACTUATION - The process that causes a switch to change position, i.e., to open, close, or transfer voltage from one contact to another. The actuating device is the mechanism that produces this process. The actuating force is the amount of pressure, movement, or other stimulus required to complete the process.

ACTUATOR - The part of a relay system that converts electrical energy into mechanical motion.

AIR GAP - The air space in the magnetic circuit of a relay and the space between contacts.

AMPERE-TURNS - The product of the number of turns in a coil and the current (in amperes) flowing through the coil.

ARMATURE - In an electromagnetic relay, the moving ferrous member that converts magnetic force into mechanical movement.

ARMATURE CHATTER - The undesired vibration of a relay armature that is due to either inadequate AC current or external shock and vibration.

ARMATURE CONTACT - A contact mounted directly on a relay armature that functions as the movable contact.

AUXILIARY CONTACTS - Contacts that operate a visual or audible signal to indicate the position of the main contacts, establish interlocking circuits, or hold a relay in a latched condition when the original operating circuit ceases to operate.

BANK - One contact level of a rotary or stepping switch.

BEZEL - The decorative ring section of a switch that is mounted on a cabinet panel.

BREAK - The opening of a closed set of contacts to interrupt an electrical circuit.

BREAK-BEFORE-MAKE CONTACT - The contact configuration of a switch or a relay that operates by opening one circuit before closing another. It is also called a "non-shorting" contact. On a relay, it is called Form C Transfer Contact.

BREAKDOWN VOLTAGE - See DIELECTRIC STRENGTH.

COIL - A number of turns of insulated wound around an iron core (or other ferrous material core) or a non-ferrous core. The wire can also be self-supporting, as in an air core

COIL POWER DISSIPATION - The electrical power (in watts) consumed by the energized winding(s) of a coil.

COIL RESISTANCE - The total terminal-to-terminal resistance of a coil at a specified temperature.

CONTACT ARC - The discharge of energy that occurs between mating contacts when the contacts are opened.

CONTACT BOUNCE - The momentary and decreasing rebounds occurring between two contact surfaces that have suddenly hit each other before a firm closureis established. Bounce time is expressed as the time interval required for reaching final closure after the initial impact.

CONTACT CONFIGURATION - The assembly of contacts in a relay.

CONTACT FORCE - The pressure exerted by a movable contact against a fixed contact when the contacts are closed.

CONTACT INTERFACE - The mating parts of two contact surfaces.

CONTACT PRESSURE - See CONTACT FORCE.

CONTACT RESISTANCE - The electrical resistance of a set of closed contacts.

CONTACT(ING) SEQUENCE - The order in which contacts open and close in relation to other contacts and armature motion.

CONTACT TRANSFER TIME - The time interval between the moving contact opening from a "break" position to the "making" with the opposite contact.

CONTACT WIPE - The scrubbing action between closing contacts resulting from the follow through of the closing action.

CROSSTALK - The electrical coupling between one set of contacts and other open or closed sets of contacts on the same relay.

DAMPER SPRING - An auxiliary spring added in some relay designs to prevent unwanted movement of a relay member in the presence of contact bounce. Damper springs are used with armatures, armature bearing pins, and contact members.

DEBOUNCER - An electronic circuit in a keyboard assembly which effectively eliminates the effects of contact bounce by maintaining closure of the key switch circuits until all bouncing has stopped.

DETENT - A small mechanism in which a spring-loaded ball or prong falls into matching indentations in a fixed plate or ring. The detent creates precise positioning of a movable contact with respect to fixed mating contacts. Also called INDEXING MECHANISM.

DIELECTRIC STRENGTH - The maximum voltage that can be applied between contacts and the frame without rupturing the insulating material.

DIP RELAY - A relay enclosed in a dual-in-line package (DIP) designed for insertion into a printed circuit board or socket.

DIP SWITCH - An assembly of several switches in a miniature housing called a dual-in-line package (DIP) designed for insertion into a printed circuit board.

DOUBLE POLE CONTACTS - A contact combination (either make, break, or transfer) where two separate sets of contacts, when actuated, operate simultaneously.

DOUBLE THROW CONTACTS - A transfer set of contacts.

DRY CIRCUIT - A circuit in which the open-circuit voltage is 50 mV (or less) and the current is 1 mA (or less).

DRY-REED CONTACTS - A glass-enclosed, magnetically operated set of contacts with metal reeds as the actuating mechanism.

DYNAMIC CONTACT RESISTANCE - The fluctuating resistance of a closed set of contacts which is due to variations in contact pressure.

ELASTOMER - A rubber-like material used in membrane keyswitches. Its significant characteristic is its memory capability that allows it to return to its original form after being depressed or flexed. It often has conductors or shorting pads printed on its underside.

ELECTRICAL OR HAND RESET - A qualifying term meaning that the contacts of a relay must be reset to their original condition (electrically or manually) following actuation of the relay.

ELECTRICAL OR MECHANICAL BIAS - An electrical or mechanical force that holds the armature of a relay toward or away from a given position until a specific actuating force is applied.

ELECTROSTATIC SHIELD - A metallic shield, or foil, usually grounded, that is placed between reed switches, between a reed switch and a coil, or between adjacent relays to minimize crosstalk.

END-OF-LIFE CRITERIA - A specified set of limiting electrical and/or mechanical conditions; any one of these conditions indicate that a switch or relay has reached the end of its useful life.

FRAME - The support structure of a relay which may also serve as part of the magnetic path.

HCL: HIGH, COMMON, LOW - A type of relay control in which a momentary contact between the common lead and a third lead causes the relay to return to its original position. This control is used with such devices as thermostats and relays.

HEADER - The sub-assembly of a relay package which supports and insulates the leads passing through the walls of a sealed relay.

HEATER - A resistance element that converts electrical energy into heat to operate a thermal relay.

HOME (HOME POSITION) - The normal, non-actuated, or starting position of a stepping relay.

INSULATION RESISTANCE - Resistance between two points of a relay when a DC voltage (100 to 500 volts) is applied.

INTERLOCK SWITCH - A switch that causes equipment in operation to de-energize when a door or other access to the equipment is opened. The purpose of this switch is to prevent malfunctioning or injury to equipment or personnel.

INTERRUPTER CONTACTS - A set of contacts on a stepping relay operated directly by an armature that opens and closes the coil circuit permitting a relay to step itself.

KEY SWITCH - A switch operated by a key. The key is removable for security purposes.

KEYBOARD SWITCH - A momentary-action, pushbutton switch with a rectangular or round key-cap. These switches are mounted in an array with terminals interconnected as a matrix. It is used for conventional typewriter-style and programming keyboards.

LATCH - A device or circuit that maintains an actuated position or condition until it is reset to its former state by external means.

LOW ENERGY SWITCH - See DRY CIRCUIT.

MAKE - The closure of a set of open contacts to actuate a circuit.

MAKE-BEFORE-BREAK CONTACTS - A set of double throw contacts arranged to close a new circuit before opening an old one. Also called a "shorting" contact on a switch or a Form D Transfer Contact or Continuity Transfer on a relay.

MANUAL ACTUATION - Operation of a switch accomplished by the action of a human.

MATING FORCE - The force that holds switch contacts in a closed (operating) state.

MERCURY-WETTED CONTACTS - A set of sealed contacts coated with a thin film of mercury rising by capillary action from a pool at the base of the contact arms of a relay.

MOMENTARY FEATURE - The characteristic of a switch that will allow it to return to its non-actuated state when the actuating force is removed. This feature is created by spring-loading the actuating mechanism.

MOVABLE CONTACT - The member of a set of contacts that is moved directly by the actuating system. It may be part of the armature or swinger spring.

MUST-OPERATE VOLTAGE - The maximum control voltage at which a relay will always be actuated. Sometimes called PICK-UP VOLTAGE.

MUST-RELEASE VOLTAGE - The minimum control voltage at which a relay switches from actuated to non-actuated. Sometimes called DROP-OUT VOLTAGE

NONBRIDGING CONTACTS - An arrangement of contacts in which the opening contact opens before the closing contact closes.

NORMAL BRIDGING - The normal make-before-break action of a make-break (Form D) contact combination in a relay.

NORMALLY-CLOSED (NC) SWITCH - A switch that maintains the closed condition of its contacts in a non-actuated state and opens its contacts when actuated.

NORMALLY-OPEN (NO) SWITCH - A switch that must be actuated to attain closure while otherwise maintaining its open state.

OPERATE TIME - The time interval (not including bounce time) between energizing a relay coil and opening of a set of normally-closed contacts, the closing of a set of normally-open contacts, or the final operation of a set of transfer contacts.

PHYSICAL ACTUATION - Actuation by a mechanical force that is not initiated by a human operator or electrical mechanism.

PILEUP - An assembly of separate sets of contacts in a relay.

POLARIZED RELAY - A relay that will operate only when the proper polarity of DC voltage is applied to the coil.

POLE - A combination of mating contacts: normally-open (NO), normally-closed (NC), or both (transfer).

POLE FACE - The part of a magnetic structure nearest the armature.

POLE PIECE - The end of an electromagnet, sometimes separable from the main section, and usually shaped to allow the distribution of the magnetic field in a pattern best suited to the application.

POWER RELAY - A relay with heavy-duty contacts that are rated at 15 amperes or more, 28 volts DC, or 115/230 volts AC, or more.

RADIO FREQUENCY INTERFERENCE (RFI) - The generation of high frequency electromagnetic waves caused by a spark or arc (the ionization of air) when the contacts of a relay are opened or closed. This condition particularly exists during the normal operation of an electromechanical relay opening and closing a high power circuit.

REED SWITCH - A pair of metal reeds mounted in an evacuated glass capsule with terminal leads brought through the capsule ends. A magnetic field produced by an external permanent magnet or an electromagnet will magnetize the reeds and cause them to attract each other. They then make contact to complete the external circuit connected to their terminals.

REPEAT KEY - A key on some keyboards that will cause any other key to repeat. With a time-delay feature, any key can be designed to repeat if held longer than a specified time. In many keyboards, this function may be furnished on all keys or selected keys and the time-delay and repeat rate can be preset. Some keyboards provide a bi-level position operation where depression of a key past the normal travel operates the repeat function.

ROCKER-ARM SWITCH - A switch actuated by a lever pivoted at its center. The lever is positioned at about a 45° angle witth respect to the panel on which the switch is mounted.

ROTARY SWITCH - A switch assembled on a central shaft and actuated by rotation of the shaft.

SENSITIVITY - The power required to operate a relay. The less power required for operation, the more sensitive the relay.

SHIELD - A magnetic structure around a relay assembly designed to keep the coil's magnetic field within its own enclosure and to prevent external magnetic fields from inadvertently actuating the relay.

SHIFT LOCK - A means of enabling a keyboard to generate two coded outputs, such as lower case and upper case letters, without holding the shift key depressed. "Secretary shift" is similar to the traditional typewriter shift. In "alternate action" shift lock, the operator depresses the shift lock key and releases it, leaving the key depressed and in the shift mode. Operating the shift lock key a second time returns the keyboard to the unshifted mode.

SLIDE SWITCH - A switch containing sets of fixed contacts and a movable shorting bar controlled by a sliding rectangular or round lever on the face of the switch.

SNAP-ACTION SWITCH - A switch with a contact actuating mechanism that provides a high velocity transfer of the moving contact from one extreme position to the other during actuation.

STEPPER RELAY - A multiposition relay in which moving wiper contacts mate with successive sets of fixed contacts in a series of steps, moving from one step to the next in successive operations of the relay. Also called STEPPING SWITCH or ROTARY STEPPING SWITCH.

STROKE - The vertical or horizontal displacement of the actuating lever which is necessary to achieve contact actuation in a switch.

SURGE CURRENT - The maximum short-term current that a closed set of contacts can conduct without damage to the contact surface, (pitting or burning) which would impair its performance.

TELEPHONE RELAY - A relay similar to the type originally developed for telephone system switching equipment.

TEMPERATURE-SENSITIVE SWITCH - A switch that is actuated when its temperature reaches a pre-set value.

THROW - The number of positions in which a switch operates.

TOGGLE SWITCH - A snap-action switch that is actuated by a projecting lever, cylinder, or handle.

TOUCH SWITCH - A switch that requires no discernible movement of its actuating button or plate to cause actuation.

WIPING CONTACTS - Contacts that provide a sliding or wiping movement at the contact faces during opening and closure.

MAGNETIC COMPONENTS

CHOKE - A component consisting of insulated wire wound around a core. A choke impedes or resists the flow of alternating current or pulsating direct current. The terms choke, coil, and inductor are generally interchangeable.

COIL - A component consisting of insulated wire wound around a ferrous or nonferrous core.

FARADAY SHIELD - A metallic foil or a network of parallel conductors connected to a common conductor that, in turn, is connected to a grounded transformer frame. The Faraday shield provides protection against electrostatic coupling without affecting the electromagnetic waves; it is part of the transformer assembly.

IMPEDANCE (Z) - The total opposition to the flow of alternating current or pulsating DC in a circuit that includes DC resistance, capacitive reactance, and inductive reactance. It is measured in ohms and designated with the capital letter "Z".

IMPEDANCE-MATCHING TRANSFORMER - A transformer that matches the output impedance of a circuit (connected to its primary) to the input impedance of a circuit (connected to its secondary).

IMPREGNATED WINDING - A winding permeated with a phenolic varnish, or other insulating material, to protect it from mechanical vibration, fungus, and moisture.

INDUCTANCE (L) - The characteristic of an inductor that produces a changing magnetic field when alternating current flows through it. The changing magnetic field produces a changing voltage in a direction opposite to the current. The more iron in the core and the more turns of wire in the winding, the greater the inductance. The unit of inductance is measured in henrys (H).

INDUCTIVE REACTANCE (X_L) - The opposition or AC resistance offered to the flow of AC or pulsating DC by an inductor. It is measured in ohms and is directly proportional to inductance and to the frequency of the applied voltage. X_L is equal to $2\pi fL$

INDUCTOR - The generic term for an electrical component that consists of insulated wire wound around a core. The term is generally interchangeable with the terms "choke" or "coil". The core can be a metallic or non-metallic hollow tube or a solid rod, and it may be filled with ferrite, powdered iron, laminated iron, steel, or air.

MAGNETIC FIELD (MAGNETIC FLUX) - The region surrounding a permanent magnet, electromagnet, or energized coil in which a magnetic force operates on magnetic (ferrous) objects in the region.

PASSIVE DELAY LINE - An inductor/capacitor circuit (LC network) that delays a signal (either linear or digital) for a predetermined time. The network requires no external power.

POWER CHOKE - A large inductor with a laminated iron or steel core that impedes the flow of low frequency (60 to 800 Hz) AC or pulsating DC. It is a part of an efficient (low loss) filter circuit in conjunction with filter capacitors to change pulsating DC into steady-state DC.

PRIMARY AND SECONDARY WINDINGS - The input (primary) and output (secondary) windings of a transformer.

RF CHOKE - A small inductor with a non-ferrous core that impedes the flow of high frequency RF current.

SATURATION - The point at which any further increase in current flowing through a magnetic component will produce no further increase in the magnetic field.

TRANSFORMER - A component that transfers electrical energy from one circuit to another, or several others, by magnetic induction, while electrically isolating the circuits from each other. It has the capability to transfer this electrical energy efficiently and, at the same time, change levels of voltage and current between and among the circuitsin a system.

TRANSFORMER TAP - A terminal of a point on a secondary winding which provides a means of connecting to a portion of the total secondary voltage. A secondary can have a single tap, a center tap, or multiple taps.

TURNS-RATIO - The ratio of the number of turns in the primary winding to the number of turns in a secondary winding. This ratio could be either step-up (a greater number of turns in the secondary) or step-down (a fewer number of turns in the secondary).

CONNECTORS

ACCORDIAN - A type of connector contact where a flat spring is given a "Z" shape to permit high deflection without stress.

BAYONET COUPLING - A quick coupling device for plug and receptacle connectors using pins on one connector and ramps on the mating device. Coupling is accomplished by rotation of the coupling ring to bring the two halves of the connector together.

BERYLLIUM COPPER - A contact alloyed material with properties superior to spring brass and phosphor bronze. It is recommended for contact applications requiring repeated extraction and reinsertion because of its resistance to fatigue at high operating temperatures.

BIFURCATED CONTACT - A connector contact slotted lengthwise to provide two independently operating points of contact. They can be used as the contacts of insulation displacement connectors (IDC).

BODY - The main portion of a connector comprised of a shell and insert to which other components are attached.

BOOT - A form placed around the wire termination of a multiple-contact connector to hold a liquid potting compound before it hardens. It serves as a protective housing since it is made from a resilient material to prevent entry of moisture into a connector.

BUSHING - A separate, high-strength metal section located in a connector to resist stress or wear. It is also a piece of insulated tubular liner used to electrically insulate coaxial conductors in a connector.

CABLE CLAMP - A mechanical clamp attached to the cable side of a connector to support the cable or wire bundle and to provide strain relief at the connection. It absorbs tension, shock, and vibration that would otherwise be transmitted to the contact/wire connection.

CARD EDGE (OR EDGE CARD) CONNECTOR - See PC BOARD CARD EDGE CONNECTOR

CARD SLOT - A lengthwise connecting section on a PC board that accepts a printed circuit board or a mating card edge connector.

CHAMFER - The angled face on the inside edge of the barrel entrance of a connector which permits easy insertion of the mating connector into the barrel.

COAXIAL CONNECTOR - Used with coaxial cable to provide high frequency bandwidth capability in communications and video systems. Coaxial cable consists of an insulated solid or stranded conductor surrounded by a solid or braided metallic shield and wrapped in insulating material such as Teflon™ or rubber.

CONNECTOR - A mechanical device that provides electrical mating between two or more sections of an electrical or electronic circuit; a pair consists of a mating plug and receptacle. Various types include: PC board edge, DIP, ribbon, circular, wire-wrapped, hermaphroditic, rack and panel, and insulation displacement connector (IDC).

CONNECTOR INSERT - Part of connectors having metal shells. The insert is the section holding contacts in proper arrangement while electrically insulating them from each other and from the shell.

CONNECTOR SHELL - The case enclosing the connector insert and connector assembly. Shells of mating connectors protect projecting contacts and provide proper alignment.

CONTACT - The conducting part of a connector that interacts with a mating part to complete a circuit.

CONTACT RESISTANCE - The electrical resistance of a pair of engaged contacts. Resistance is measured at a specified current flowing through the engaged contacts; it is expressed in ohms or millivolts divided by the specified current.

CONTACT RETAINER - A device placed either on a contact or in an insert to secure the position of the contact.

CONTACT SHOULDER - The flanged portion of the contact which limits its travel into the insert.

CONTACT SPACING - The distance between the centers of contacts within an insert.

CONTACT WIPE - The length of travel made by one contact touching another during the mating and unmating of a connector assembly. Also called WIPING ACTION.

CRIMP - The act of physically compressing or deforming a connector or contact barrel around a cable to make an electrical connection.

CUT-OUT - A round or rectangular hole cut in a metal panel for mounting a single-contact or multiple-contact connector.

DEGREASE - An operation used in the manufacture of connectors to remove unwanted oils or grease deposited on the connectors or other components during production.

EGG-CRATING - The insulating walls between the cavities within the contact or wire entry face of the connector housing. Egg-crating allows the contact to be fully protected by insulating material to prevent shorts and to minimize shock.

FERRULE - A short tube used to make solderless connections to a shielded or coaxial cable.

FLANGE - A projection extending from or around the periphery of a connector. It is provided with holes to permit mounting the connector to a panel or to a mating connector.

FLAT CABLE CONNECTOR - A connector designed for a flat-conductor flat cable or a round-conductor flat cable.

FRAME - The surrounding metal portion of a multiple-contact connector having a removable body or insert. The frame supports the insert and provides a method for mounting the connector to a panel or mating connector.

GROMMET - A rubber seal located on the cable side of a connector to seal it against moisture, dirt, or air.

HOUSING - A connector part without an insert, but equipped with insert-retaining and positioning hardware.

INSULATION DISPLACEMENT CONNECTOR (IDC) - A mass termination connector for flat or ribbon cable with contacts that displace the conductor insulation and establish separate connections to each conductor.

KEY - A short pin or other projection that slides into a mating slot or groove to guide two parts being assembled. Generally used in round, shell-enclosed, or card edge connectors to guarantee proper orientation of the mating sections.

O-RING - A doughnut-shaped rubber ring used as a seal around the periphery of the mating insulator interface of cylindrical connectors.

PC BOARD CARD EDGE CONNECTOR - A connector that mates with printed circuit conductors running to the edge of a printed circuit board. Also called CARD EDGE CONNECTOR.

PHOSPHOR BRONZE - An alloy of copper, tin, and phosphorus used in the manufacture of contact springs in connectors. It is harder than spring brass and is used in applications where the temperature is 105°C or less. Phosphor bronze is reliable as long as insertion and withdrawal operations are not too frequent.

PIN CONTACT - A male type contact designed to mate with a socket or female contact. It is normally connected to the "dead side" of a circuit. The "dead side" is defined as contacts that are not powered.

PLUG - The part of the two mating halves of a connector pair which is free to move when not fastened to the other mating half. The plug is usually, but not always, the male portion of the pair. The plug may have female contacts if it is the "free to move" member.

POLARIZATION - A mechanical arrangement of inserts and/or shell configuration that prohibits the mating of mismatched plugs and receptacles. Polarization allows connectors of the same size to be aligned side by side with no danger of making the wrong connection. Different arrangements of contacts, keys, keyways, and insert positions are used.

QUICK DISCONNECT - A type of connector shell that permits rapid locking and unlocking of two connectors.

RECEPTACLE - The fixed half of a connector pair usually mounted on a panel or chassis. The receptacle is normally, but not necessarily, the female contact portion of the connector pair.

SOCKET CONTACT - A female contact, normally connected to the "live side" of a circuit, which will mate with a male contact.

SOLDER-EYE TERMINAL - A solder type terminal with a hole at its end through which a conductor is inserted prior to being soldered.

SOLDERLESS CONNECTION - The joining of two conductors by pressure without soldering, brazing, or employing other methods requiring heat.

SOLDERLESS WRAP - A method of connecting a solid wire to a square, rectangular, or V-shaped terminal by using a special tool to tightly wrap the wire around the terminal. Also called WIREWRAP.

TNC SERIES CONNECTOR - A radio frequency connector with an impedance of 50 ohms designed to operate in the 10 gigahertz range. It has a threaded coupling to prevent an accidental disconnect.

APPENDIX

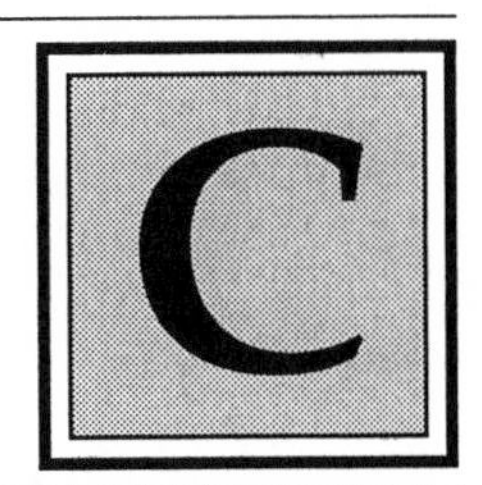

GLOSSARY OF ELECTRONIC SYMBOLS

GLOSSARY OF ELECTRONIC SYMBOLS

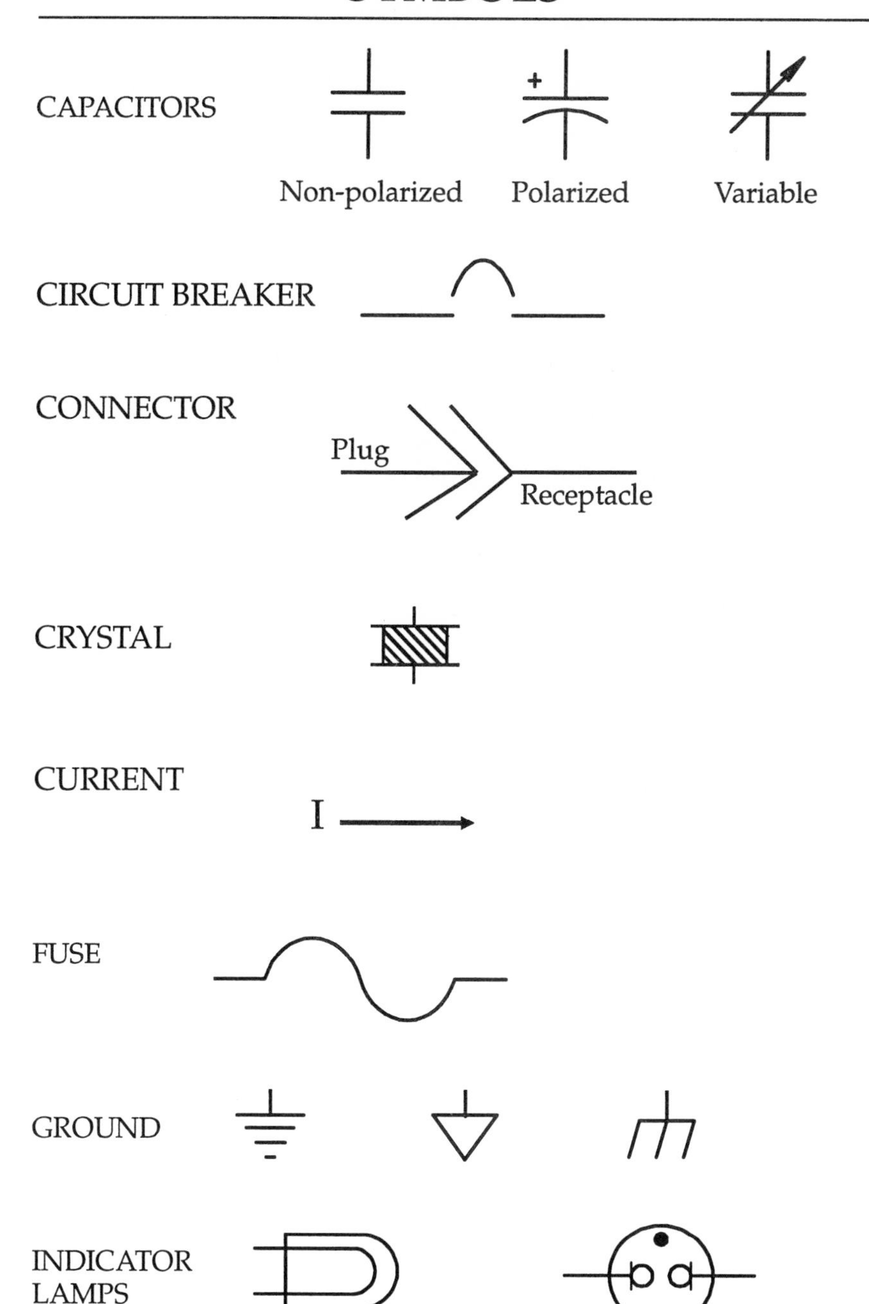

INDUCTORS

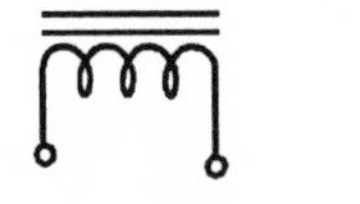

Laminated Iron or Steel Core

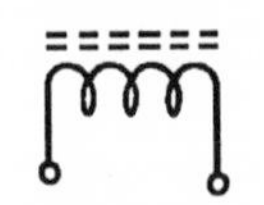

Ferrite Core (Powdered iron)

Non-ferrous or Air Core

METERS

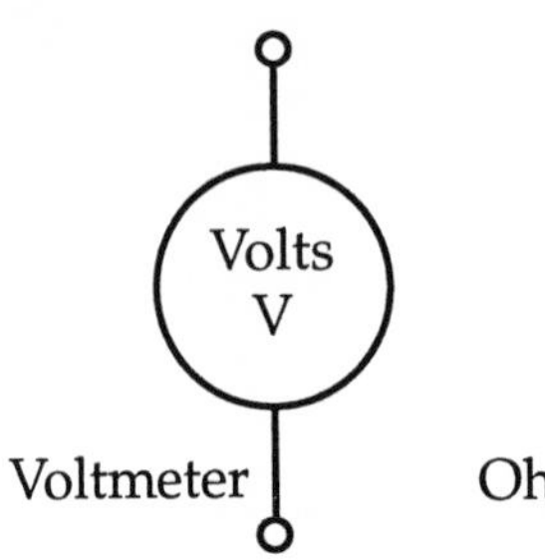

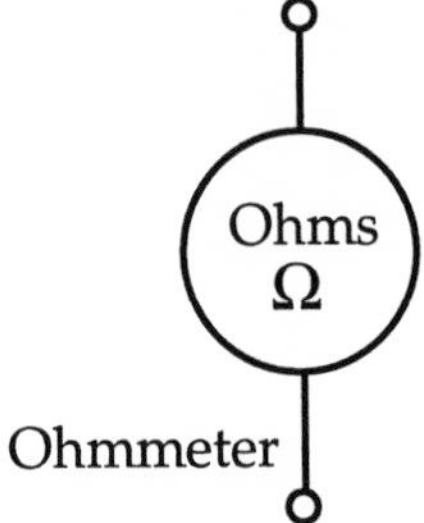

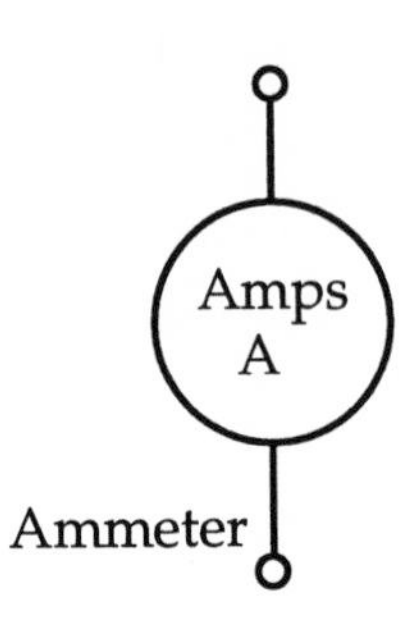

RELAY (ELECTROMECHANICAL)

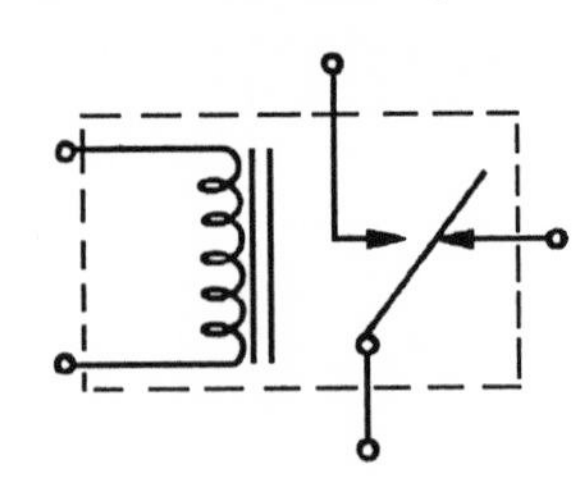

RESISTOR (FIXED)

RESISTORS (VARIABLE)

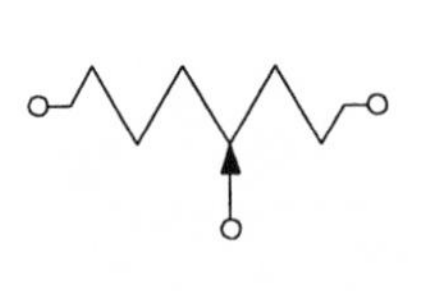

Potentiometer

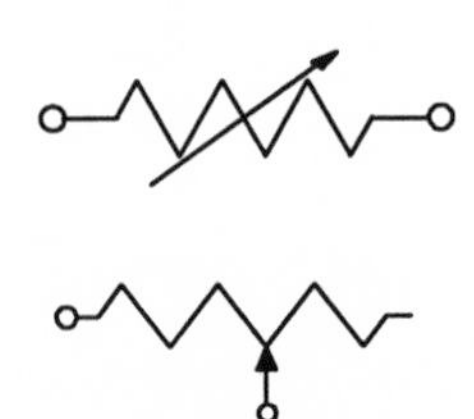

Rheostats

SWITCHES (MECHANICAL)

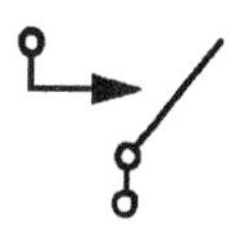

Single Pole, Single Throw (SPST) Normally Open (NO)

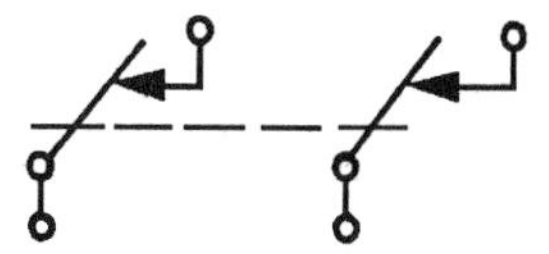

Double Pole, Single Throw (DPST) Normally Closed (NC)

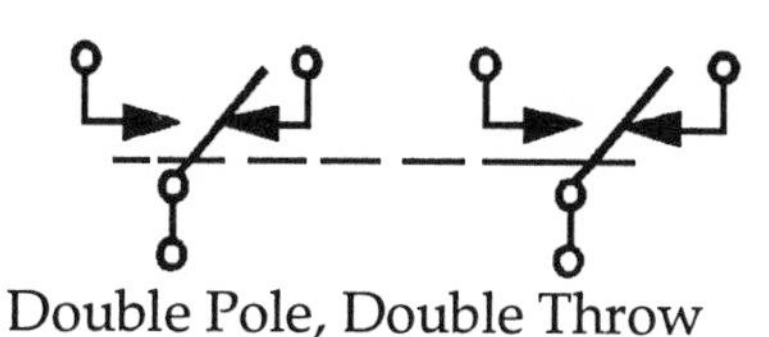

Double Pole, Double Throw (DPDT)

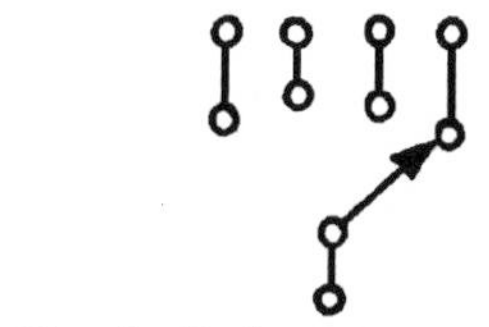

Single Pole, 4 Position (SP, 4POS)

TRANSFORMERS

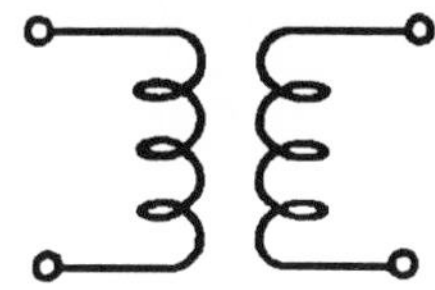

Non-Ferrous or Air Core

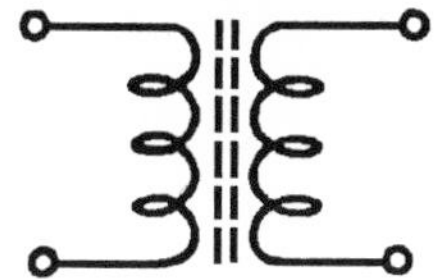

Ferrite Core (Powdered iron)

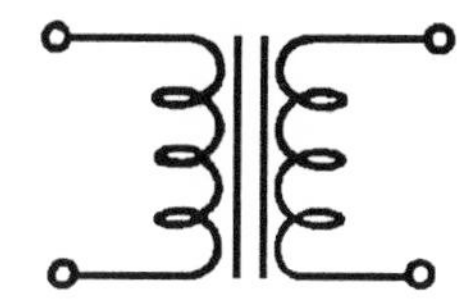

Laminated Iron or Steel

VOLTAGE SOURCE (SUPPLY VOLTAGE) - AC

Fixed

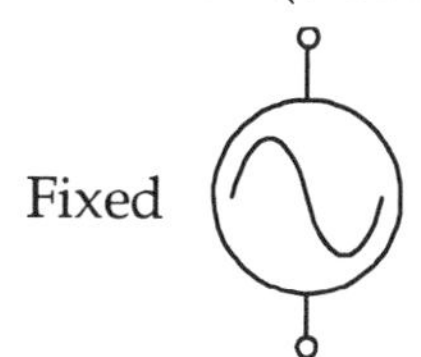

Variable

VOLTAGE SOURCE (SUPPLY VOLTAGE) - DC

Fixed

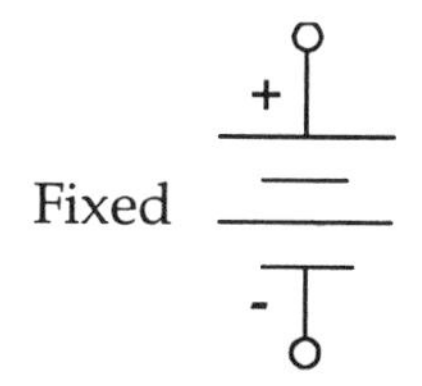

Variable

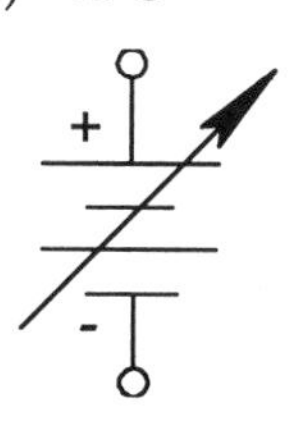

REFERENCES

Asimov, Isaac. *ASIMOV'S BIOGRAPHICAL ENCYCLOPEDIA OF SCIENCE AND TECHNOLOGY.* Second Rev. ed.
Garden City, N.Y.: Doubleday & Company, Inc., 1982

Cheney, Margaret. *TESLA, MAN OUT OF TIME.*
Englewood Cliffs, N.J.: Prentice-Hall, Inc., 1981

Douglas-Young, John. *ILLUSTRATED ENCYCLOPEDIC DICTIONARY OF ELECTRONICS.*
West Nyack, N.Y.: Parker Publishing Company, Inc., 1981

Hausmann, Erich and Slack, Edgar P. *PHYSICS.* Second ed.
New York, N.Y.: D. Van Nostrand Company, 1939

Graf, Rudolf. *MODERN DICTIONARY OF ELECTRONICS.*
Fourth ed. Indianapolis, Indiana: Howard W. Sams, Inc., 1972

Klein, Herbert Arthur. THE SCIENCE OF MEASUREMENT - A Historical Survey
Mineola, N.Y.: Dover Publications, 1988

Levine, Sy. A LIBRARY ON BASIC ELECTRONICS -
Volume Two - Discrete Semiconductors & Optoelectronics;
Volume Three - Integrated Circuits and Computer Concepts
Plainview, N.Y.: Electro-Horizons Publications

Smith, F. Langford, editor. *RADIOTRON DESIGNER'S HANDBOOK.* Fourth ed.
Harrison, N.J.: Radio Corporation of America, 1953

ELECTRONIC BUYER'S HANDBOOK & DIRECTORY, 1987.
Manhasset, N.Y.: CMP Publications, Inc.

ELECTRONIC ENGINEERS MASTER (EEM) 1995, 4 vols.
Garden City, N.Y.: Hearst Business Communications, Inc.

INDEX

A

B

C

D

E

F

G

H

I

J

K

L

M

N

O

P

Q

R

S

T

U

V

W

X

Z